Robotic Non-Destructive Testing

Robotic Non-Destructive Testing

Editors

Carmelo Mineo
Yashar Javadi

MDPI • Basel • Beijing • Wuhan • Barcelona • Belgrade • Manchester • Tokyo • Cluj • Tianjin

Editors
Carmelo Mineo
National Research Council (CNR)
Italy

Yashar Javadi
University of Strathclyde
UK

Editorial Office
MDPI
St. Alban-Anlage 66
4052 Basel, Switzerland

This is a reprint of articles from the Special Issue published online in the open access journal *Sensors* (ISSN 1424-8220) (available at: https://www.mdpi.com/journal/sensors/special_issues/robotic_NDT).

For citation purposes, cite each article independently as indicated on the article page online and as indicated below:

LastName, A.A.; LastName, B.B.; LastName, C.C. Article Title. *Journal Name* **Year**, *Volume Number*, Page Range.

ISBN 978-3-0365-6529-3 (Hbk)
ISBN 978-3-0365-6530-9 (PDF)

© 2023 by the authors. Articles in this book are Open Access and distributed under the Creative Commons Attribution (CC BY) license, which allows users to download, copy and build upon published articles, as long as the author and publisher are properly credited, which ensures maximum dissemination and a wider impact of our publications.
The book as a whole is distributed by MDPI under the terms and conditions of the Creative Commons license CC BY-NC-ND.

Contents

About the Editors . . . . **vii**

Carmelo Mineo and Yashar Javadi
Robotic Non-Destructive Testing
Reprinted from: *Sensors* 2022, 22, 7654, doi:10.3390/s22197654 1

Gong Zhang, Yuhang Zhang, Shuaihua Tuo, Zhicheng Hou, Wenlin Yang, Zheng Xu, et al.
A Novel Seam Tracking Technique with a Four-Step Method and Experimental Investigation of Robotic Welding Oriented to Complex Welding Seam
Reprinted from: *Sensors* 2021, 21, 3067, doi:10.3390/s21093067 5

**Momchil Vasilev, Charles N. MacLeod, Charalampos Loukas, Yashar Javadi,
Randika K. W. Vithanage, David Lines, et al.**
Sensor-Enabled Multi-Robot System for Automated Welding and In-Process Ultrasonic NDE
Reprinted from: *Sensors* 2021, 21, 5077, doi:10.3390/s21155077 19

**Rastislav Zimermann, Ehsan Mohseni, Momchil Vasilev, Charalampos Loukas,
Randika K. W. Vithanage, Charles N. Macleod, et al.**
Collaborative Robotic Wire + Arc Additive Manufacture and Sensor-Enabled In-Process Ultrasonic Non-Destructive Evaluation
Reprinted from: *Sensors* 2022, 22, 4203, doi:10.3390/s22114203 35

Alastair Poole, Mark Sutcliffe, Gareth Pierce and Anthony Gachagan
A Novel Complete-Surface-Finding Algorithm for Online Surface Scanning with Limited View Sensors
Reprinted from: *Sensors* 2021, 21, 7692, doi:10.3390/s21227692 49

**Euan Alexander Foster, Gary Bolton, Robert Bernard, Martin McInnes, Shaun McKnight,
Ewan Nicolson, et al.**
Automated Real-Time Eddy Current Array Inspection of Nuclear Assets
Reprinted from: *Sensors* 2022, 22, 6036, doi:10.3390/s22166036 65

Minglu Zhang, Xuan Zhang, Manhong Li, Jian Cao and Zhexuan Huang
Optimization Design and Flexible Detection Method of a Surface Adaptation Wall-Climbing Robot with Multisensor Integration for Petrochemical Tanks
Reprinted from: *Sensors* 2020, 20, 6651, doi:10.3390/s20226651 85

F. Javier Garcia Rubiales, Pablo Ramon Soria, Begoña C. Arrue and Anibal Ollero
Soft-Tentacle Gripper for Pipe Crawling to Inspect Industrial Facilities Using UAVs
Reprinted from: *Sensors* 2021, 21, 4142, doi:10.3390/s21124142 105

**Jiaxing Wang, Mingqiang Yang, Zhongjun Ding, Qinghe Zheng, Deqiang Wang,
Kidiyo Kpalma and Jinchang Ren**
Detection of the Deep-Sea Plankton Community in Marine Ecosystem with Underwater Robotic Platform
Reprinted from: *Sensors* 2021, 21, 6720, doi:10.3390/s21206720 125

**Kamil Cetin, Harun Tugal, Yvan Petillot, Matthew Dunnigan, Leonard Newbrook
and Mustafa Suphi Erden**
A Robotic Experimental Setup with a Stewart Platform to Emulate Underwater Vehicle-Manipulator Systems
Reprinted from: *Sensors* 2022, 22, 5827, doi:10.3390/s22155827 143

Carmelo Mineo, Nicola Montinaro, Mario Fustaino, Antonio Pantano and Donatella Cerniglia
Fine Alignment of Thermographic Images for Robotic Inspection of Parts with Complex Geometries
Reprinted from: *Sensors* **2022**, *22*, 6267, doi:10.3390/s22166267 . . . . **159**

Nicolas P. Avdelidis, Antonios Tsourdos, Pasquale Lafiosca, Richard Plaster, Anna Plaster and Mark Droznika
Defects Recognition Algorithm Development from Visual UAV Inspections
Reprinted from: *Sensors* **2022**, *22*, 4682, doi:10.3390/s22134682 . . . . **179**

About the Editors

Carmelo Mineo

Carmelo Mineo (PhD) received a master's degree in Mechanical Engineering from the University of Palermo (Italy) in 2011. In 2012, he joined the Centre for Ultrasonic Engineering of the University of Strathclyde (Glasgow, UK) to undertake his doctoral studies on the automated, non-destructive inspection of large and complex geometries of composite materials. He became a Research Associate of the University of Strathclyde in 2015 and a Research Fellow in 2018. Carmelo was awarded a prestigious H2020 Marie-Curie Fellowship in 2020, funded by the European Commission, to lead research on Robotic Adaptive Behaviors for NDT Inspections in Dynamic Contexts at the University of Palermo. He has been a Researcher at the Institute of High-Performance Computing and Networking of the National Research Council of Italy since 2020. His current research interests comprise intelligent and autonomous robotics, advanced robot control for real-time adaptive path-planning, instrument and sensor interfacing, data collection and processing.

Yashar Javadi

Yashar Javadi (PhD) is a Strathclyde Chancellor's Fellow (Lecturer as part of Strathclyde Global Talent Programme) at the University of Strathclyde. His multidisciplinary works have been reflected in his appointment, which is a joint appointment between two departments (Department of Electronic & Electrical Engineering and Department of Design Manufacturing & Engineering Management). In a career spanning over 16 years in the field of engineering (with a particular focus on wire-arc welding, non-destructive evaluation, additive manufacturing, and robotics), he has worked as a lecturer and postdoctoral research associate (at The University of Manchester and the University of Strathclyde) in academia and as a welding engineer and manager of welding/NDT departments in the industry. At present, he is a member of Centre for Ultrasonic Engineering, where his research focus is on the in-process inspection and residual stress measurement (non-destructive methods) in robotic welding and additive manufacturing applications. His interests comprise the renewable energy and structural integrity of the offshore wind turbine.

Editorial

Robotic Non-Destructive Testing

Carmelo Mineo [1,*] and Yashar Javadi [2,3]

1. Institute for High-Performance Computing and Networking, National Research Council, 153, 90146 Palermo, Italy
2. Department of Electronic & Electrical Engineering, University of Strathclyde, Glasgow G1 1XW, UK
3. Department of Design, Manufacturing & Engineering Management, University of Strathclyde, Glasgow G1 1XW, UK
* Correspondence: carmelo.mineo@icar.cnr.it

1. Introduction

Non-destructive testing (NDT) and evaluation (NDE) are commonly referred to as the vast group of analysis techniques used in civil, medical, and industrial sectors to evaluate the properties of materials, tissues, components, or structures without causing any damage. NDT/NDE are vital to ensure the integrity of critical parts and social safety. Automation offers many benefits for NDT to cope with increasing demands, including improved reliability and higher inspection speeds. Additionally, robots enable inspection positions that are not easily accessible to human operators to be reached and enable humans to be removed from potentially dangerous environments. However, the perceived complexity and high costs have limited the adoption of automation for NDT. As a result, the full potential that could be derived from the seamless integration of robotic platforms with sensors, actuators, and software has not been fully explored; it could still revolutionise the way automated NDT is performed and conceived. Robots are often operated by predefined tool paths generated through offline path-planning software applications. The recent advancements in electronics, robotics, sensor technology and software pave the way for new developments in automated NDT and data-driven autonomous robotic inspections in several sectors. This Special Issue aimed to attract the latest research outcomes in the field of robotic NDT. It comprises eleven high-quality papers. Five papers relate to inspection systems based on robotic fixed-base manipulators. Three research articles are associated with in-process inspection in manufacturing applications (robotic wire-arc welding and additive manufacturing). Four papers report research advancements in mobile robotic-enabled sensing. The remaining two papers focus on novel developments in data visualisation and analysis.

2. Overview of Contribution

Among the five papers related to inspection systems based on robotic fixed-base manipulators, three are associated with in-process inspection in robotic wire-arc welding and additive manufacturing. Zhang et al. [1] introduce a novel seam tracking technique for robotic welding. The method is proven effective, providing a reference for future seam tracking research. Vasilev et al. [2] present the development and deployment of an advanced multi-robot system for automated welding and in-process NDE. Complete external positional control is achieved in real time, allowing on-the-fly motion correction based on multi-sensory input. This approach can enable in-process weld repair, leading to higher production efficiency, reduced rework rates, and lower production costs. Zimermann et al. [3] introduce a synchronised multi-robot Wire + Arc Additive Manufacturing and NDE cell aiming to achieve in-process defect detection, enable possible in-process repair, and prevent the costly scrappage or reworking of completed defective builds. A novel high-temperature-capable, dry-coupled phased array ultrasound transducer roller-probe

device is used for the NDE inspection. The dry-coupled sensor is tailored for coupling with an as-built high-temperature surface at an applied force and speed.

The other two papers on fixed-base robotic arms introduce a novel algorithm for complete-surface-finding [4] and an automated system for the real-time eddy current inspection of nuclear assets [5]. The work presented in the first article enables the robot-assisted ultrasonic testing of unknown surfaces within a single pass, a significant advancement toward fully autonomous inspection systems. The latter paper introduces a system capable of delivering an eddy current array to detect stress corrosion cracking on a nuclear canister. The variation in the lift-off of the eddy current array is innovatively minimised using a force–torque sensor, a padded flexible probe, and a feedback control system.

Four papers focus on mobile robotic applications. Zhang et al. [6] introduce a flexible design and defect detection method for a multi-sensor, wall-climbing robot used to inspect petrochemical tanks. The results show that the robot can move safely and stably on a vertical tank surface and complete precise automatic detection. Rubiales et al. [7] present a crawling mechanism using a soft-tentacle gripper integrated into an unmanned aerial vehicle for pipe inspection in industrial environments. The objective was to allow the aerial robot to perch and crawl along a pipe, minimising energy consumption and performing contact inspection. This paper introduces the design of the soft limbs of the gripper and the internal mechanism that allows movement along pipes. The other two papers in this group relate to underwater robotic inspection applications. Wang et al. [8] present a novel method for use in deep-sea plankton community detection in marine ecosystems using an underwater robotic platform. This paper demonstrates that moving plankton can be accurately detected and isolated from complex dynamic backgrounds in deep-sea environments. Cetin et al. [9] present an experimental robotic setup with a Stewart platform and a robot manipulator to emulate an underwater vehicle–manipulator system. The hardware-based emulator setup consists of a robotic manipulator mounted on a parallel manipulator, known as a Stewart Platform, and a force–torque sensor attached to the end-effector of the robotic arm interacting with a pipe. Such a complete setup is useful to use when carrying out fast and numerous experiments, circumventing the difficulties in performing similar experiments and data collection with actual underwater vehicles in water tanks.

Robotic inspection systems can acquire substantial data volumes. As a result, the research into robotic NDT/NDE must also embrace some efforts to introduce new data visualisation and analysis approaches. That is the case with the two remaining papers of this Special Issue. Mineo et al. [10] introduce an image alignment method to facilitate the visualisation and analysis of robotic thermographic inspections of parts with complex geometries. This work bridges a technology gap, making thermographic inspections more deployable in industrial environments. The proposed image alignment approach can find applicability beyond thermographic non-destructive testing. Finally, Avdelidis et al. [11] propose a two-step process for the automation of defect recognition and classification from visual images. This can be used with unmanned aerial vehicles carrying an image sensor to automate the procedure and eliminate human error.

3. Conclusions

Robotic NDT is a fast-evolving field which exploits the constant advancements in electronics, robotics, sensor technology, software, and network interfaces. This Special Issue is a collection of eleven high-quality publications that provide a picture of some of the most commonly investigated topics in robotic-enabled sensing. In-process inspection in robotic manufacturing applications, real-time and data-driven robotic sensing, and mobile terrestrial, underwater, and aerial robotic inspection platforms are well represented. The authors have also proposed innovative solutions to improve the visualisation and analysis of large robotically collected datasets. The advancements in robotic NDT help us face new societal challenges, which the industrial sectors have encapsulated under so-called Industry 4.0. Robotic NDT must develop with the development of new tools, including autonomous robotics, virtual-twin simulations, the Internet of Things, cybersecurity, cloud computing, augmented reality, and big data. For this reason, this Special Issue is not the conclusion of a path but the prelude to upcoming collections of research outcomes.

Funding: This research received no external funding.

Institutional Review Board Statement: Not applicable.

Informed Consent Statement: Not applicable.

Data Availability Statement: Not applicable.

Conflicts of Interest: The authors declare no conflict of interest.

References

1. Zhang, G.; Zhang, Y.; Tuo, S.; Hou, Z.; Yang, W.; Xu, Z.; Wu, Y.; Yuan, H.; Shin, K. A Novel Seam Tracking Technique with a Four-Step Method and Experimental Investigation of Robotic Welding Oriented to Complex Welding Seam. *Sensors* **2021**, *21*, 3067. [CrossRef] [PubMed]
2. Vasilev, M.; MacLeod, C.; Loukas, C.; Javadi, Y.; Vithanage, R.; Lines, D.; Mohseni, E.; Pierce, S.; Gachagan, A. Sensor-Enabled Multi-Robot System for Automated Welding and in-Process Ultrasonic NDE. *Sensors* **2021**, *21*, 5077. [CrossRef] [PubMed]
3. Zimermann, R.; Mohseni, E.; Vasilev, M.; Loukas, C.; Vithanage, R.K.W.; Macleod, C.N.; Lines, D.; Javadi, Y.; E Silva, M.P.E.; Fitzpatrick, S.; et al. Collaborative Robotic Wire + Arc Additive Manufacture and Sensor-Enabled in-Process Ultrasonic Non-Destructive Evaluation. *Sensors* **2022**, *22*, 4203. [CrossRef] [PubMed]
4. Poole, A.; Sutcliffe, M.; Pierce, G.; Gachagan, A. A Novel Complete-Surface-Finding Algorithm for Online Surface Scanning with Limited View Sensors. *Sensors* **2021**, *21*, 7692. [CrossRef] [PubMed]
5. Foster, E.A.; Bolton, G.; Bernard, R.; McInnes, M.; McKnight, S.; Nicolson, E.; Loukas, C.; Vasilev, M.; Lines, D.; Mohseni, E.; et al. Automated Real-Time Eddy Current Array Inspection of Nuclear Assets. *Sensors* **2022**, *22*, 6036. [CrossRef] [PubMed]
6. Zhang, M.; Zhang, X.; Li, M.; Cao, J.; Huang, Z. Optimization Design and Flexible Detection Method of a Surface Adaptation Wall-Climbing Robot with Multisensor Integration for Petrochemical Tanks. *Sensors* **2020**, *20*, 6651. [CrossRef] [PubMed]
7. Rubiales, F.G.; Soria, P.R.; Arrue, B.; Ollero, A. Soft-Tentacle Gripper for Pipe Crawling to Inspect Industrial Facilities Using UAVs. *Sensors* **2021**, *21*, 4142. [CrossRef] [PubMed]
8. Wang, J.; Yang, M.; Ding, Z.; Zheng, Q.; Wang, D.; Kpalma, K.; Ren, J. Detection of the Deep-Sea Plankton Community in Marine Ecosystem with Underwater Robotic Platform. *Sensors* **2021**, *21*, 6720. [CrossRef] [PubMed]
9. Cetin, K.; Tugal, H.; Petillot, Y.; Dunnigan, M.; Newbrook, L.; Erden, M.S. A Robotic Experimental Setup with a Stewart Platform to Emulate Underwater Vehicle-Manipulator Systems. *Sensors* **2022**, *22*, 5827. [CrossRef] [PubMed]
10. Mineo, C.; Montinaro, N.; Fustaino, M.; Pantano, A.; Cerniglia, D. Fine Alignment of Thermographic Images for Robotic Inspection of Parts with Complex Geometries. *Sensors* **2022**, *22*, 6267. [CrossRef] [PubMed]
11. Avdelidis, N.P.; Tsourdos, A.; Lafiosca, P.; Plaster, R.; Plaster, A.; Droznika, M. Defects Recognition Algorithm Development from Visual UAV Inspections. *Sensors* **2022**, *22*, 4682. [CrossRef] [PubMed]

Article

A Novel Seam Tracking Technique with a Four-Step Method and Experimental Investigation of Robotic Welding Oriented to Complex Welding Seam

Gong Zhang [1,2], Yuhang Zhang [1,3], Shuaihua Tuo [1,3], Zhicheng Hou [1,*], Wenlin Yang [1], Zheng Xu [1], Yueyu Wu [1], Hai Yuan [1] and Kyoosik Shin [4]

1 Guangzhou Institute of Advanced Technology, Chinese Academy of Sciences, Guangzhou 511458, China; gong.zhang@giat.ac.cn (G.Z.); yh.zhang@giat.ac.cn (Y.Z.); sh.tuo@giat.ac.cn (S.T.); wl.yang@giat.ac.cn (W.Y.); zheng.xu@giat.ac.cn (Z.X.); yy.wu@giat.ac.cn (Y.W.); hai.yuan@giat.ac.cn (H.Y.)
2 School of Engineering Science, University of Chinese Academy of Sciences, Beijing 100049, China
3 School of Construction Machinery, Chang'an University, Xi'an 710064, China
4 Department of Robot Engineering, ERICA Campus, Hanyang University, Seoul 426-791, Korea; norwalk87@hanyang.ac.kr
* Correspondence: zc.hou@giat.ac.cn

Citation: Zhang, G.; Zhang, Y.; Tuo, S.; Hou, Z.; Yang, W.; Xu, Z.; Wu, Y.; Yuan, H.; Shin, K. A Novel Seam Tracking Technique with a Four-Step Method and Experimental Investigation of Robotic Welding Oriented to Complex Welding Seam. Sensors 2021, 21, 3067. https://doi.org/10.3390/s21093067

Academic Editors: Carmelo Mineo and Yashar Javadi

Received: 1 April 2021
Accepted: 25 April 2021
Published: 28 April 2021

Publisher's Note: MDPI stays neutral with regard to jurisdictional claims in published maps and institutional affiliations.

Copyright: © 2021 by the authors. Licensee MDPI, Basel, Switzerland. This article is an open access article distributed under the terms and conditions of the Creative Commons Attribution (CC BY) license (https://creativecommons.org/licenses/by/4.0/).

Abstract: The seam tracking operation is essential for extracting welding seam characteristics which can instruct the motion of a welding robot along the welding seam path. The chief tasks for seam tracking would be divided into three partitions. First, starting and ending points detection, then, weld edge detection, followed by joint width measurement, and, lastly, welding path position determination with respect to welding robot co-ordinate frame. A novel seam tracking technique with a four-step method is introduced. A laser sensor is used to scan grooves to obtain profile data, and the data are processed by a filtering algorithm to smooth the noise. The second derivative algorithm is proposed to initially position the feature points, and then linear fitting is performed to achieve precise positioning. The groove data are transformed into the robot's welding path through sensor pose calibration, which could realize real-time seam tracking. Experimental demonstration was carried out to verify the tracking effect of both straight and curved welding seams. Results show that the average deviations in the X direction are about 0.628 mm and 0.736 mm during the initial positioning of feature points. After precise positioning, the average deviations are reduced to 0.387 mm and 0.429 mm. These promising results show that the tracking errors are decreased by up to 38.38% and 41.71%, respectively. Moreover, the average deviations in both X and Z direction of both straight and curved welding seams are no more than 0.5 mm, after precise positioning. Therefore, the proposed seam tracking method with four steps is feasible and effective, and provides a reference for future seam tracking research.

Keywords: welding robot; seam tracking; laser sensor; feature point extracting; complex welding seam

1. Introduction

Mechanical robots have become crucial for modern welding owing to high-volume profitability since manual welding yields low production rates [1]. Robotic welding brings different favorable circumstances, for instance, it has made strides in efficiency, weld quality, adaptability and workspace use, and it diminishes work costs in addition to focused unit cost [2].

Be that as it may, most welding robots still work in the working mode of "teach and playback" and their adaptability is not enough when the welding object or other conditions are changed [3]. Since welding as an empirical process is influenced by numerous factors, such as the mistakes of pre-machining, fitting of work pieces, and in-process defects, can result in variation in welding seam. However, welding robots in teach and playback mode have no such capacities and typically weld a weldment with many defects and poor penetration [1].

There are generally three stages in robotic welding: (i) preparation—calibration, robot programming, and weld parameter, work-piece setting, (ii) welding—seam tracking, alternation of weld parameters in real time, (iii) analysis—weld quality inspection [4]. The seam tracking operation is essential for extracting weld seam characteristics which can be fed into the controller of welding robot to instruct the motion of the robot along the welding seam path. Seam tracking technology with laser vision sensing has the advantages of no contact, fast speed, and high precision, which are the keys to realizing welding automation and intelligence [5,6].

In order to fulfill the required welding accuracy for robotic welding, a seam tracking algorithm that enables the robot to plan its path along the actual welding line is necessary. Therefore, many studies have been conducted on automatic seam tracking using sensors such as tactile, touch, probe, vision sensors [7,8], laser sensors [9,10], arc sensors [11,12], electromagnetic sensors [13,14], and ultrasonic sensors [15,16]. The sensors have a very important role in robotic seam tracking; the chief tasks would be weld starting and ending points detection, weld edge detection, joint width measurement.

A basic laser sensor consists of three parts: laser diode, CCD camera, and filter. The laser diode could produce a stripe or dot which would be scanned by the camera. The CCD camera is always fixed at an angle to the laser to capture properly the projection of laser on the work piece [17]. The welding seam tracking system based on laser vision combines laser measurement and computer vision technology. It has the advantages of rich information acquisition, obvious welding seam characteristics, and strong anti-interference ability [18,19], which are suitable for real-time tracking systems. The mathematical model of transforming the laser feature points pixel coordinate to the three-dimensional coordinate of the welding feature points by designing the mechanical structure of the sensor was proposed [20].

Chen et al. [21] proposed a feature points positioning method that only needs two profile scans, which can effectively calculate the initial position of the weld. Chang et al. [22] filtered, derived and convolved the weld profile data, and located the feature points by finding the local maxima. Wang et al. [23] established welding seam profile detection and feature points extracting algorithms based on a NURBS-snake and visual attention model, and verified their effectiveness. Mastui et al. [24] introduced an adaptive welding robot system controlled by laser sensor for welding of thin plates with gap variation in single pass.

In a flexible welding process, Ciszak et al. [25] developed a low-cost system for identifying shapes in order to program industrial robots for a welding process in two dimension. The programming of industrial robots was to detect geometric shapes proposed by humans and to approximate them. Based on this, the robot could weld the same profiles on a two-dimensional plane. This is time-consuming as many welding robot applications are programmed by teach and playback, which means that they need to be reprogrammed each time they deal with a new task. Hairol et al. [26] suggested an alternative approach that can automatically recognize and locate the butt-welding position at starting, middle, auxiliary, and end point under three conditions which are (i) straight, (ii) saw tooth, and (iii) curve joint. This was done without any prior knowledge of the shapes involved. As an automatic welding process may experience different disturbances, Li et al. [27] proposed a robust method for identifying this seam based on cross-modal perception so as to precisely identify and automatically track the welding seam.

Wojciechowski et al. [28] proposed the method of automatic robotic assembly of two or more parts placed without fixing instrumentation and positioning on the pallet, which could support a robotic assembly process based on data from optical 3D scanners. The sequence of operations from scanning to place the parts in the installation position by an industrial robot was developed. Suszynski et al. [29] presented the concept of using an industrial robot equipped with a triangulation scanner in the assembly process in order to minimize the number of clamps that could hold the units in a particular position in space based on the proposed multistep processing algorithm.

These efforts have brought about many improvements in the feature points of the target weldment. However, there are certain limitations in the positioning accuracy due the factors such as the change of the welding type (especially oriented to complex welding seam) or the surface defects of the welding.

Due to these circumstances, we here introduce a novel seam tracking technique with a four-step method. First, a laser sensor is used to scan the groove of the weldment to collect profile data; then the data are processed by a filtering algorithm to smooth the noise; next, the second derivative algorithm is proposed to initially locate the feature points based on linear fitting to accurately locate the feature points; finally, according to the results of the sensor pose calibration, the three-dimensional coordinates in the base coordinate system of the welding robot are calculated from the two-dimensional coordinates of the image feature points, and the path planning is completed, with both the line and curve of the Y-shaped groove being targeted as well. The proposed seam tracking technique is tested and verified by way of experimental investigation.

Our proposed seam tracking technique with a four-step method utilizes edge detection and curvature recognition techniques based on laser scan data. The offset of the welding robot's motion with respect to the welding seam is measured by a laser sensor. By adding a differential point searching method, the feature points of the cross-section of the welding seam are found. Comparing to other seam tracking algorithms, we show the improvement of the required welding accuracy oriented to complex welding seam through theoretical proof, simulation, and experiments.

This paper is organized as follows: Section 2 presents the seam tracking system composition; Section 3 introduces the seam tracking methodology with four steps; Section 4 shows the results of the experimental investigation based on the proposed seam tracking technique; Section 5 gives the conclusion and perspective.

2. Seam Tracking System Composition

The experimental platform composition of the six-axis robot arm for seam tracking system is detailed in Figure 1. As evident in Figure 1, this experimental platform is mainly composed of the motion execution mechanism with six degrees of freedom, laser vision sensor, D/A conversion module, and industrial computer, robotic controller, welding equipment, i.e., welding power supply and wire feeding device, etc.

The execution mechanism is composed of two welding robots, and each of them has six degrees of freedom. The offset of the welding robot's motion with respect to the welding seam is measured by a laser vision sensor. Through robotic welding experiments, images of molten pool morphology and welding geometry under different welding parameters can be obtained. The main tasks for seam tracking would be weld starting and ending point detection, weld edge detection, joint width measurement, and weld path position determination with regard to welding robot co-ordinate frame.

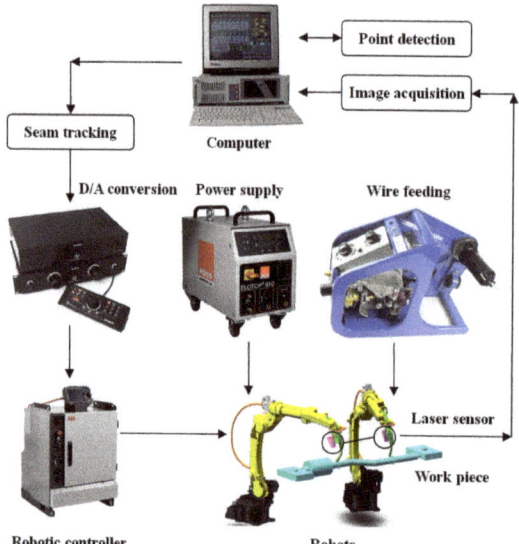

Figure 1. Diagram of seam tracking system.

3. Seam Tracking Methodology with Four Steps

In this paper, we introduce a novel seam tracking technique with a four-step method: scanning, filtering, feature points extracting, and path planning. Firstly, the profile information is obtained by scanning the groove with a laser sensor; then, the data are filtered to smooth the noise; next, the feature points are extracted by the combination of the second derivative algorithm and linear fitting; finally, the data of the feature points are converted into the welding seam path of the robot, guiding the welding torch to move and realize the real-time tracking of the welding seam. The flowchart of the proposed four-step method is revealed in Figure 2.

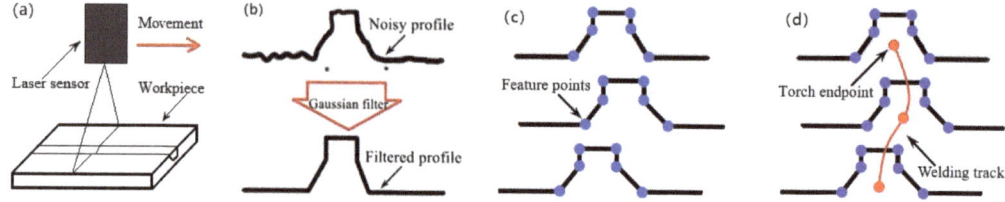

Figure 2. Flowchart of the four-step method for (**a**) scanning; (**b**) filtering; (**c**) feature points extracting; and (**d**) path planning.

3.1. Scanning and Filtering

The purpose of scanning is to obtain the original data of the weldment groove profile, which is the basis for realizing seam tracking [30]. The laser sensor obtains the distance information of the measured object based on the principle of triangulation and then processes the scan data to obtain the profile feature of the measured object. While scanning, the sensor is fixed at the end-effector of the robot and parallel to the welding torch to ensure that the line laser is perpendicular to the measured object [31], covering the groove to the greatest extent, and at the same time, the welding robot is constantly moved to obtain the overall shape of the welding seam.

The combination of limiting filter and Gaussian filter is used to process the groove profile data obtained by scanning. The former is used to remove the pulse interference caused by accidental factors. The latter is used to smooth the data [32]. The data are processed using limiting filtering by comparing the absolute value of the difference between two adjacent sample values and the size of the threshold. Its principle can be expressed as [33]:

$$y = \begin{cases} y_n & |y_n - y_{n-1}| \leq \Delta T \\ y_{n-1} & |y_n - y_{n-1}| > \Delta T \end{cases}, \quad (1)$$

where y_n and y_{n-1} are the current and last sampled signal values, respectively, and ΔT represents the specified threshold.

Gaussian filtering is a type of linear smoothing filtering method that selects weights according to the shape of the Gaussian function. It is very effective in suppressing the noise that obeys the normal distribution [34], and the Gaussian function has good properties of symmetry, differentiability, and integrability. The function can accurately identify the discontinuous points of the signal, which is very beneficial for the subsequent feature points extracting. The expression of the one-dimensional Gaussian function can be described as [35]:

$$f(x) = \frac{1}{\sigma\sqrt{2\pi}} e^{-\frac{(x-\mu)^2}{2\sigma^2}}, \quad (2)$$

where μ is the mean value, which determines the position of the function, and σ is the standard deviation, which determines the magnitude of the distribution.

3.2. Feature Point Extracting

The feature points of the weldment are generally the corner points of the groove section, and its information can reflect the overall situation of the groove profile [36], so feature point extracting is required. This is done according to the cross-sectional characteristics of the weldment groove, combined with the related properties of the function discontinuities listed in Table 1. The groove feature points could be classified as follows: **A, B, E, F**, which are the first type of feature points, and **C, D**, which are the second type of feature points, as shown in Figure 3.

Table 1. Properties of discontinuous points of function.

Discontinuous Points Type	Amplitude	First Derivative	Second Derivative
The first	continuity	Step mutation	extremum
The second	continuity	non-existent	/

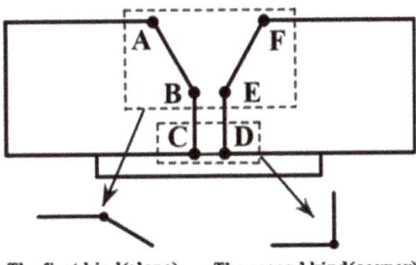

Figure 3. Classification of groove feature points.

Based on the above analysis, the feature points can be located by determining the types of feature points contained in the groove section, and then deriving them to find the extreme points.

3.2.1. Initial Positioning of Feature Points

The preliminary positioning method of the groove feature points is as follows: First, the original data are processed by filtering, and then the first derivative is obtained by the forward difference method and the extreme points are found to determine the first type of feature points, as compared in Figure 4. The abscissa and the ordinate, respectively, represent the X and Z axes of the sensor coordinate system.

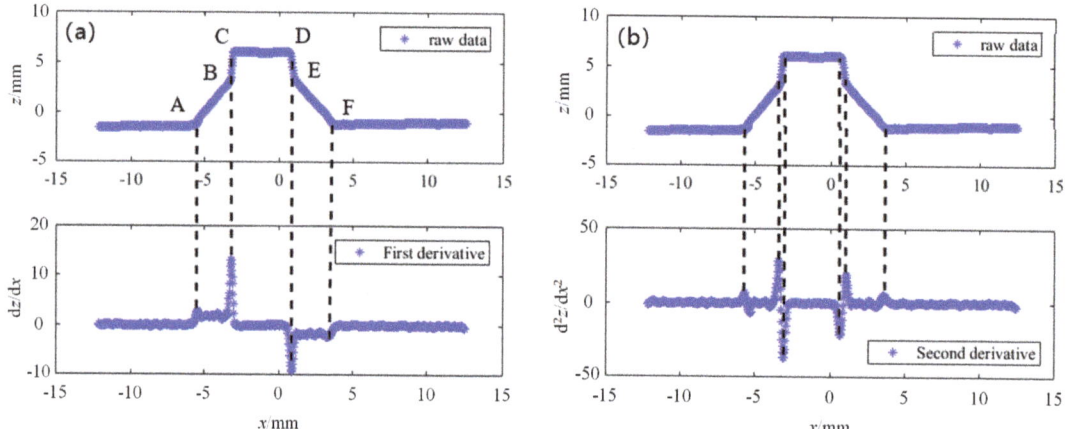

Figure 4. Initial positioning of feature points for (**a**) the first type of feature points; and (**b**) all feature points.

It can be seen from the above figures that the maximum point of the first-order guide falls between the line segment **BC** and **DE**, and fails to accurately correspond to **B** and **E**. This is because the groove of the weldment under actual conditions needs to be machined, and its blunt edge is not a vertical line in an ideal state, but a diagonal line. Therefore, the second type of feature points are transformed into the first type, and the first-order derivative can be continued to differ, and the second-order derivative can be obtained and the point with the highest value can be found to locate all the feature points, as shown in Figure 4. So far, the six characteristic points of the trapezoidal groove have been preliminarily determined, and their location information is listed in Table 2.

Table 2. Results of initial positioning.

Feature Points	A	B	C	D	E	F
X/mm	−5.67	−3.37	−3.02	0.72	1.11	3.59
Z/mm	−1.35	2.89	6.03	6.01	3.15	−1.02

3.2.2. Precise Positioning of Feature Points

Due to the defects on the surface of the weldment, as given in Figure 5, the feature points obtained through preliminary positioning are **b** and **c**, while the true feature point should be **a**, which is clearly a deviation. Therefore, on the basis of preliminary positioning, linear fitting is performed on each segment of the groove to accurately locate the feature points.

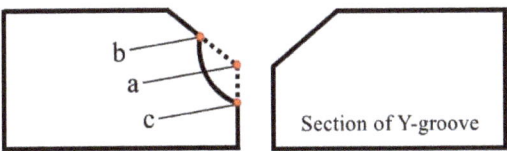

Figure 5. Defects on the surface of the weldment.

Suppose any straight-line equation to be fitted is $y = ax + b$, and the calculation of equation parameters can be written as [37]:

$$\begin{bmatrix} a \\ b \end{bmatrix} = \begin{bmatrix} \sum_{i=1}^{n} x_i^2 & \sum_{i=1}^{n} x_i \\ \sum_{i=1}^{n} x_i & n \end{bmatrix}^{-1} \cdot \begin{bmatrix} \sum_{i=1}^{n} x_i y_i \\ \sum_{i=1}^{n} y_i \end{bmatrix}, y = \begin{cases} y_n & |y_n - y_{n-1}| \leq \Delta T \\ y_{n-1} & |y_n - y_{n-1}| > \Delta T \end{cases} \quad (3)$$

where a is the slope, b is the intercept, (x_i, y_i) is the point passing through the straight line, and n is the number of points.

The fitting results are shown in Figure 6, and the relevant parameters of the straight line are illustrated in Table 3.

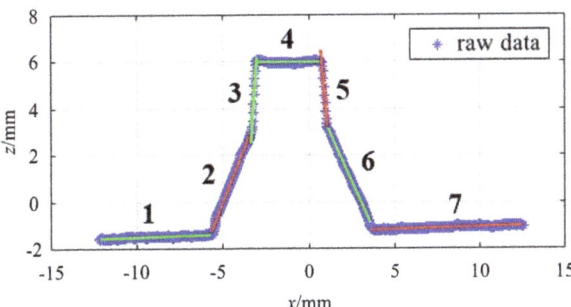

Figure 6. Fitting results.

Table 3. Parameters of fitting straight line.

Fitting Straight Line	1	2	3	4	5	6	7
SSE	0.08	0.44	0.39	0.15	0.50	0.15	0.21
R-squared	0.85	0.99	0.95	0.87	0.97	0.99	0.81

Among them, SSE is the sum variance, which calculates the sum of squared errors between the fitting data and the corresponding points of the original data. The smaller the value, the better the fitting affects; R-squared is the coefficient of determination, which is used to characterize the quality of the fitting [38]; the closer its value is to 1, the better the fitting affects. It is easy to know that the fitting effect of each straight line is better. The results of precise positioning of the feature points are listed in Table 4. So far, the feature points extracting of the profile for the trapezoidal groove section would be completed.

Table 4. Results of precise positioning.

Feature Points	A	B	C	D	E	F
X/mm	−5.73	−3.31	−3.04	0.78	1.10	3.76
Z/mm	−1.39	3.07	5..98	5..99	3.22	−1.18

3.3. Path Planning

Because the data measured by the laser sensor are based on their own coordinate system, it is necessary to convert the feature points to the base coordinate system of the welding robot through pose calibration [39].

The relationship between two coordinate systems of the robot is depicted in Figure 7. The sensor calibration is to determine the transformation matrix ${}^{E}_{S}T$ of $\{S\}$ relative to $\{E\}$.

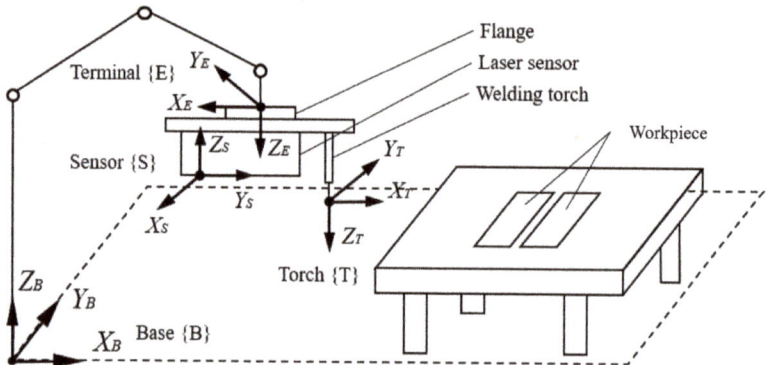

Figure 7. Relationship between two coordinate systems.

This paper uses the multipoint method for calibration [40]. The main steps are as follows:

1. Select a point P on the weldment, make the end of the welding torch this point, and record the position of P in the $\{B\}$ coordinate system ${}^{B}P = (x_B, y_B, z_B, 1)^T$, as shown in Figure 8a.
2. Move the robot so that the laser line of the sensor passes through this point, and record the position of P in the $\{S\}$ coordinate system ${}^{S}P = (x_S, 0\ z_S, 1)^T$, as shown in Figure 8b.
3. Switch the current tool coordinate system of the robot to $\{E\}$, record the pose data of the robot at this time, and from the Euler rotation equation, ${}^{B}_{E}R$ can be expressed as [41]:

$$
{}^{B}_{E}R = \begin{bmatrix} \cos\alpha & -\sin\alpha & 0 \\ \sin\alpha & \cos\alpha & 0 \\ 0 & 0 & 1 \end{bmatrix} \cdot \begin{bmatrix} \cos\beta & 0 & \sin\beta \\ 0 & 1 & 0 \\ -\sin\beta & 0 & \cos\beta \end{bmatrix} \cdot \begin{bmatrix} 1 & 0 & 0 \\ 0 & \cos\gamma & -\sin\gamma \\ 0 & \sin\gamma & \cos\gamma \end{bmatrix} = \begin{bmatrix} R_{11} & R_{12} & R_{13} \\ R_{21} & R_{22} & R_{23} \\ R_{31} & R_{32} & R_{33} \end{bmatrix}, \quad (4)
$$

where α, β, γ are the rotation angles of the X, Y, and Z axes of the tool coordinate system $\{E\}$, respectively.

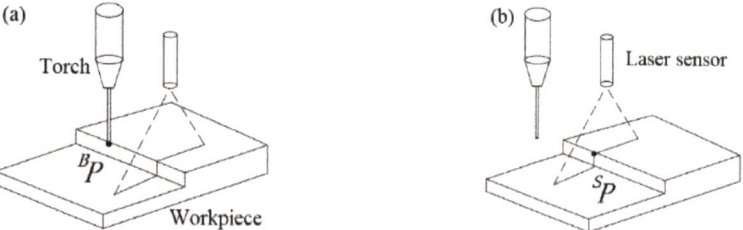

Figure 8. Laser sensor calibration for (a) base coordinates; and (b) sensor coordinates.

Then, $_E^BT$ can be simplified to

$$_E^BT = \begin{bmatrix} _E^BR & ^EP \\ 0\ 0\ 0 & 1 \end{bmatrix}, \qquad (5)$$

where $^EP = (x_E, y_E, z_E)^T$, that is, the position of point P in the tool coordinate system $\{E\}$ after the coordinate system is switched.

According to the transformation relationship of point P in space:

$$^BP = {_E^BT} \cdot {_S^ET} \cdot {^SP}, \qquad (6)$$

where the definition of each parameter in the formula is consistent with the above.

Since $_S^ET$ contains 12 unknowns, at least 3 different fixed points need to be selected to solve the problem. The calibration results in this paper are as follows:

$$_S^ET = \begin{bmatrix} 0.998 & -0.423 & -0.590 & 75.098 \\ -0.014 & 0.278 & -0.026 & 6.693 \\ 0.002 & 0.865 & -0.814 & 303.131 \\ 0 & 0 & 0 & 1 \end{bmatrix}, \qquad (7)$$

At this point, the pose calibration of the sensor is completed. For any known points ^SQ in its coordinate system, the formula to transform it into the robot base coordinate system can be written as

$$^BQ = {_E^BT} \cdot {_S^ET} \cdot {^SQ}, \qquad (8)$$

where ^BQ and ^SQ are respectively the position of point Q in the coordinate system $\{B\}$ and the coordinate system $\{S\}$; $_S^ET$ is the calibration result of Equation (4); the definition and calculation of $_S^ET$ follow step 3.

4. Experimental Procedures

Experimental demonstration had been carried out at the proposed seam tracking method with four steps to guide the movement of the welding torch under actual testing conditions. Figure 9 reveals the prototype of whole experimental system, which mainly includes ABB IRB 1410 welding robot, IRC5 controller, LS-100CN laser sensor, Ehave CM350 welding power supply, RS-485 communication module, and an industrial computer.

Figure 9. A prototype of the experimental system.

In this paper, two typical weldments with materials of A304 stainless steel are selected as the welding objects, the physical prototypes of two typical welding grooves are illustrated in Figure 10, and the groove parameters of the weldment with straight line and curve are listed in Table 5.

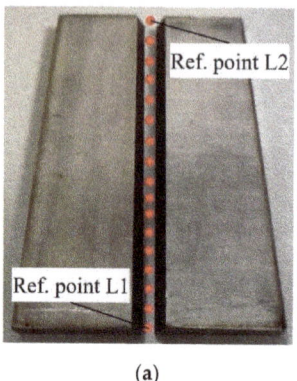 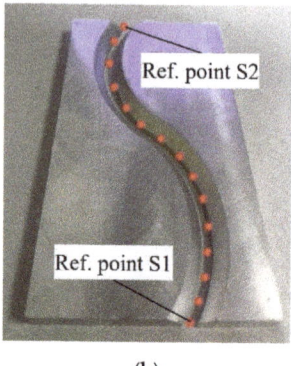

(a)   (b)

Figure 10. Two typical welding grooves for (**a**) straight line; and (**b**) curve.

Table 5. Groove parameters of weldment.

Welding Type	Dimension/mm	Thickness/mm	Slope Angle/°	Blunt Edge/mm
Straight line	100 × 60	8	45	2.5
Curve	130 × 70	10	60	3

When scanning the welding groove, the laser sensor is set to the trigger mode, and the welding robot is constantly moved to obtain the overall shape characteristics of the welding seam. The process of scanning two typical welding grooves by the laser sensor is represented in Figure 11.

 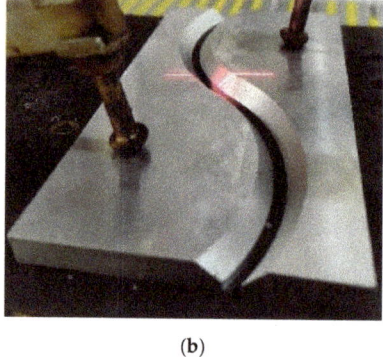

(a)   (b)

Figure 11. Two typical welding grooves scanned by laser sensor: (**a**) straight line; (**b**) curve.

Before the experiment, we mark the starting and ending points of the welding path on the weldment, and then the straight and curved grooves are respectively taught a section of motion trajectory in the model of "teach", as shown in Figure 10. The red point is the teaching point, which is the position of the end point of the robotic welding torch. Multiple teaching points are connected to form a welding trajectory, and the pose data of the teaching trajectory in the welding torch coordinate system will be recorded simultaneously, which is used as a reference to calculate the experimental deviation.

During the experiment, if the straight groove is taken as an example, let us first move the end-effector of the robot, i.e., the welding torch, along the teaching trajectory. When it reaches reference point **L1**, as shown in Figure 10a, the laser sensor will be turned on to scan the welding groove and collect data. At the same time, the current tool coordinate system of the welding robot will be switched to the end coordinate system, the position and posture data of the end coordinate system are obtained in real time through the API interface of the welding robot, and the sampling period is consistent with that of the laser sensor.

The welding robot continues to move. When the end of the welding torch moves to reference point **L2**, as shown in Figure 10a, the laser sensor will be turned off, the data transmission of the API interface is stopped, the data collection is completed. According to the feature points of the groove, the center point of the welding torch is calculated; according to the position and posture data of the end coordinate system obtained by API interface, the trajectory reference point is calculated. Through the calibration matrix of laser sensor (Formula (7)), the position data of the welding torch center point is transformed into the welding robot end coordinate system, and then through the calibration matrix of welding torch, it is transformed into the welding torch coordinate system.

After the above process, the groove data collected by the laser sensor are transformed into the center point data of the robotic welding torch, and the end coordinate system data collected by the API interface are transformed into the trajectory reference point data. The experimental results of two different welding grooves of straight and curved lines with both initial positioning and precise positioning using the proposed seam tracking method are compared in Figure 12.

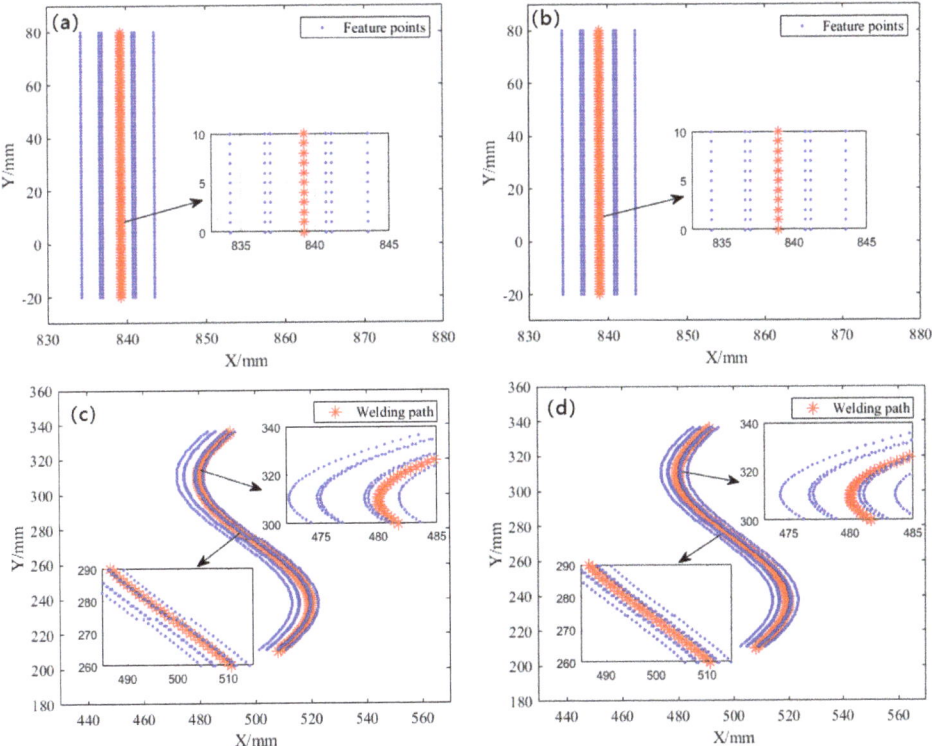

Figure 12. Experimental results of (**a**) straight line with initial positioning; (**b**) straight line with precise positioning; (**c**) curve with initial positioning; and (**d**) curve with precise positioning.

The accuracy of the feature points positioning method is evaluated by comparing the deviation between the calculated welding center point and the actual welding torch end point. Among them, the average deviation d (mm) represents the average value of the difference between each welding center point and the end point of the welding torch;

the deviation degree p (%) indicates the deviation degree of the deviation in this direction relative to the entire groove. The average deviation d (mm) and deviation degree p (%) can be written as:

$$d_x = \frac{1}{n}\sum_{i=1}^{n}\left(x_{tcp(i)} - x_{t(i)}\right), \quad d_z = \frac{1}{n}\sum_{i=1}^{n}\left(x_{tcp(i)} - z_{t(i)}\right), \quad (9)$$

where d_x and d_z are the average deviation in the X and Z directions, respectively. $x_{tcp(i)}$ and $z_{tcp(i)}$ are the coordinates of the welding center point, $x_{t(i)}$ and $z_{t(i)}$ are the coordinates of the trajectory reference point, respectively. n is the number of points.

$$p_x = \frac{d_x}{l}, \quad p_z = \frac{d_z}{h}, \quad (10)$$

where p_x and p_z are the deviation degrees the in X and Z directions, respectively. l is the total length of the groove, and h is the depth of the groove.

The comparative results of different positioning methods for feature points are depicted in Table 6. As can be seen from the figures and table, the average deviations d_x (mm) of the two different welding seams of both straight line and curve in the X direction are relatively large when only initial positioning is carried out. After precise positioning, the average deviations are reduced to 0.387 mm and 0.429 mm, respectively. Experimental procedures show promising results, in that the average deviations display a significant decrease by 38.38% and 41.71%, respectively.

Table 6. Error analysis results.

Welding Type	Initial Positioning				Precise Positioning			
	d_x/mm	d_z/mm	p_x/%	p_z/%	d_x/mm	d_z/mm	p_x/%	p_z/%
Straight line	0.628	0.214	6.688	2.665	0.387	0.230	4.121	2.864
Curve	0.736	0.185	7.838	2.304	0.429	0.251	4.569	3.126

It is worth noting that the average deviations in both X and Z direction of two different welding seams of both straight line and curve after precise positioning are no more than 0.5 mm; this value is defined by Kovacevic et al. [42] and could fulfill the minimum accuracy requirements of robotic welding. Therefore, it is suggested that the proposed seam tracking method with four steps is feasible and effective, and provides a reference for future seam tracking research.

5. Conclusions

A novel seam tracking technique and experimental investigation of robotic welding oriented to complex welding seam are proposed in this study. Conclusions are as follows:

- A set of seam tracking systems based on laser sensing and visual information extraction is designed, and the method involving scanning, filtering, feature points extracting, and path planning is proposed to realize high-precision seam tracking;
- The groove information is collected through the laser sensor and the data are filtered, and the corresponding three-dimensional coordinate value in the sensor coordinate system is calculated using the two-dimensional coordinates of the image feature points;
- The accuracy problem of feature point positioning when the weldment surface has defects is solved. Experimental results show that the average deviations of both straight line and curve of welding feature points after precise positioning is less than 0.5 mm;
- The experimental errors are mainly caused by the calibration error of the sensor coordinate system and the calculation error of the feature points extracting algorithm. In addition, increasing the resolution of the sensor could further improve the measurement accuracy.

Author Contributions: Conceptualization, G.Z. and Z.H.; methodology, G.Z. and S.T.; software, S.T., Y.Z., and Y.W.; validation, S.T. and Y.Z.; formal analysis, S.T., Z.X.; investigation, G.Z. and Z.H.; resources, S.T. and W.Y.; data curation, Y.Z., S.T., Y.W., and W.Y.; writing—original draft, S.T. and Y.Z.; writing—review and editing, G.Z., Z.H. and K.S.; visualization, S.T.; supervision, G.Z. and Z.H.; project administration, G.Z. and H.Y.; funding acquisition, G.Z. and Z.H. All authors have read and agreed to the published version of the manuscript.

Funding: This research was funded in part by the National Key Research and Development Project of China, grant number 2018YFA0902903, the National Natural Science Foundation of China, grant number 62073092, the Natural Science Foundation of Guangdong Province, grant number 2021A1515012638, the Basic Research Program of Guangzhou City of China, grant number 202002030320.

Institutional Review Board Statement: Not applicable.

Informed Consent Statement: Not applicable.

Data Availability Statement: The data presented in this study are openly available in [A Novel Seam Tracking Technique with A Four-Step Method and Experimental Investigation of Robotic Welding Oriented to Complex Welding Seam—research data] at [https://cloud.huawei.com/home#/collection/v2/all] (accessed on 15 April 2021).

Acknowledgments: The authors would like to express their thanks to the Guangzhou Institute of Advanced Technology, Chinese Academy of Sciences, for helping them with the experimental characterization.

Conflicts of Interest: The authors declare no conflict of interest.

References

1. Amruta Rout, B.B.V.L.; Deepak, B.B. Advances in weld seam tracking techniques for robotic welding: A review. *Robot. Comput. Integr. Manuf.* **2019**, *56*, 12–37. [CrossRef]
2. Pires, A.J.N.; Loureiro, T.; Godinho, P.; Ferreira, B.; Fernando, J.M. Welding robots. *IEEE Robot. Autom. Mag.* **2003**, *10*, 45–55. [CrossRef]
3. Shao, W.J.; Huang, Y.; Zhang, Y. A novel weld seam detection method for space weld seam of narrow butt joint in laser welding. *Opt. Laser Technol.* **2018**, *99*, 39–51. [CrossRef]
4. Pires, A.J.N.; Loureiro, G.B. *Welding Robots: Technology, System Issues and Application*; Springer Science & Business Media: Berlin/Heidelberg, Germany, 2006.
5. Lei, T.; Rong, Y.M.; Wang, H.; Huang, Y.; Li, M. A review of vision-aided robotic welding. *Comput. Ind.* **2020**, *123*, 103326–103355. [CrossRef]
6. Hong, L.; Xiaoqi, C. Laser visual sensing for seam tracking in robotic arc welding of titanium alloys. *Int. J. Adv. Manuf. Technol.* **2005**, *26*, 1012–1017.
7. Peiquan, X.; Guoxiang, X.; Xinhua, T.; Shun, Y. A visual seam tracking system for robotic arc welding. *Int. J. Adv. Manuf. Technol.* **2008**, *37*, 70–75.
8. Shi, F.; Tao, L.; Chen, S. Efficient weld seam detection for robotic welding based on local image processing. *Ind. Robot. Int. J.* **2009**, *56*, 277–283. [CrossRef]
9. Mikael, F.; Gunnar, B. Design and validation of a universal 6d seam tracking system in robotic welding based on laser scanning. *Ind. Robot. Int. J.* **2003**, *30*, 437–448.
10. Wu, Q.-Q.; Lee, J.-P.; Park, M.-H.; Park, C.-K.; Kim, I.-S. A study on development of optimal noise filter algorithm for laser vision system in GMA welding. *Procedia Eng.* **2014**, *97*, 819–827. [CrossRef]
11. Jeong, S.-K.; Lee, G.-Y.; Lee, W.-K.; Kim, S.-B. Development of high speed rotating arc sensor and seam tracking controller for welding robots. In Proceedings of the 2001 IEEE International Symposium on Industrial Electronics, (Cat. No.01TH8570), Pusan, Korea, 12–16 June 2001; pp. 845–850.
12. Ushio, M.; Mao, W. Modelling of an arc sensor for dc mig/mag welding in open arc mode: Study of improvement of sensitivity and reliability of arc sensors in GMA welding. *Weld. Int.* **1996**, *10*, 622–631. [CrossRef]
13. You, B.-H.; Kim, J.-W. A study on an automatic seam tracking system by using an electromagnetic sensor for sheet metal arc welding of butt joints, In: Proceedings of the institution of mechanical engineers. *Part B J. Eng. Manuf.* **2002**, *216*, 911–920. [CrossRef]
14. Kang-Yul, B.; Jin-Hyun, P. A study on development of inductive sensor for automatic weld seam tracking. *J. Mater. Process. Technol.* **2006**, *176*, 111–116.
15. Freire, B.T.; Miguel, M.J.; Leopoldo, C.; Ramdn, C. Weld seams detection and recognition for robotic arc-welding through ultrasonic sensors. In Proceedings of the 1994 IEEE International Symposium on Industrial Electronics (ISIE'94), Santiago, Chile, 25–27 May 1994; pp. 310–315.

16. Maqueira, B.; Umeagukwu, C.I.; Jarzynski, J. Application of ultrasonic sensors to robotic seam tracking. *IEEE Trans. Robot. Autom.* **1989**, *5*, 337–344. [CrossRef]
17. He, Y.; Chen, Y.; Xu, Y.; Huang, Y.; Chen, S. Autonomous detection of weld seam profiles via a model of saliency-based visual attention for robotic arc welding. *J. Intell. Robot. Syst.* **2016**, *81*, 395–402. [CrossRef]
18. Guo, J.C.; Zhu, Z.; Yu, Y.; Sun, B. Research and Application of Visual Sensing Technology Based on Laser Structured Light in Welding Industry. *Chin. J. Lasers* **2017**, *44*, 7–16.
19. Hou, Z.; Xu, Y.L.; Xiao, R.Q.; Chen, S.B. A teaching-free welding method based on laser visual sensing system in robotic GMAW. *Int. J. Adv. Manuf. Technol.* **2020**, *109*, 1755–1774. [CrossRef]
20. Zou, Y.B.; Wang, Y.B.; Zhou, W.L. Research on Line Laser Seam Tracking Method based on Guassian Kernelized Correlation Filters. *Appl. Laser* **2016**, *36*, 578–584.
21. Chen, X.H.; Dharmawan, A.G.; Foong, S.H.; Soh, G.S. Seam tracking of large pipe structures for an agile robotic welding system mounted on scaffold structures. *Robot. Comput. Integr. Manuf.* **2018**, *50*, 242–255. [CrossRef]
22. Chang, D.Y.; Son, D.H.; Lee, J.W.; Kim, T.W.; Lee, K.Y.; Kim, J.W. A new seam-tracking algorithm through characteristic-point detection for a portable welding robot. *Robot. Comput. Integr. Manuf.* **2012**, *28*, 1–13. [CrossRef]
23. Wang, N.F.; Zhong, K.F.; Shi, X.D.; Zhang, X.M. A robust weld seam recognition method under heavy noise based on structured-light vision. *Robot. Comput. Integr. Manuf.* **2020**, *61*, 1–9. [CrossRef]
24. Matsui, S.; Goktug, G. Slit laser sensor guided real-time seam tracking arc welding robot system for non-uniform joint gaps, Industrial Technology. *Proc. IEEE Int. Conf. Ind. Technol.* **2002**, *1*, 159–162.
25. Olaf, C.; Jakub, J.; Suszynski, M. Programming of Industrial Robots Using the Recognition of Geometric Signs in Flexible Welding Process. *Symmetry* **2020**, *12*, 1429.
26. Shah, H.N.M.; Sulaiman, M.; Shukor, A.Z.; Kamis, Z.; Rahman, A.A. Butt welding joints recognition and location identification by using local thresholding. *Robot. Comput. Integr. Manuf.* **2018**, *51*, 181–188. [CrossRef]
27. Xinde, L.; Pei, L.; Omar, K.M.; Xiangheng, H.; Sam, G.S. A welding seam identification method based on cross-modal perception. *Ind. Robot. Int. J. Robot. Res. Appl.* **2019**, *46*, 453–459.
28. Jakub, W.; Marcin, S. Optical scanner assisted robotic assembly. *Assem. Autom.* **2017**, *37*, 434–441.
29. Marcin, S.; Jakub, W.; Jan, Z. No Clamp Robotic Assembly with Use of Point Cloud Data from Low-Cost Triangulation Scanner. *Teh. Vjesn. Tech. Gaz.* **2018**, *25*, 904–909.
30. Zhou, G.H.; Xu, G.C.; Gu, X.P.; Liu, J.; Tian, Y.K.; Zhou, L. Simulation and experimental study on the quality evaluation of laser welds based on ultrasonic test. *Int. J. Adv. Manuf. Technol.* **2017**, *93*, 3897–3906. [CrossRef]
31. Yang, G.W.; Yan, S.M.; Wang, Y.Z. V-Shaped Seam Tracking Based on Particle Filter with Histogram of Oriented Gradient. *Chin. J. Lasers* **2020**, *47*, 330–338.
32. He, Y.S.; Yu, Z.H.; Li, J.; Yu, L.S.; Ma, G.H. Discerning Weld Seam Profiles from Strong Arc Background for the Robotic Automated Welding Process via Visual Attention Features. *Chin. J. Mech. Eng.* **2020**, *33*, 799–816. [CrossRef]
33. Zou, Y.B.; Wang, Y.B.; Zhou, W.L.; Chen, X.Z. Real-time seam tracking control system based on line laser visions. *Opt. Laser Technol.* **2018**, *103*, 182–192. [CrossRef]
34. Shao, W.J.; Liu, X.F.; Wu, Z.J. A robust weld seam detection method based on particle filter for laser welding by using a passive vision sensor. *Int. J. Adv. Manuf. Technol.* **2019**, *104*, 2971–2980. [CrossRef]
35. Chen, W.J. *Research on Seam Track Measuring System Based on Stripe Type Laser Sensor [Dissertation]*; South China University of Technology: Guangzhou, China, 2018.
36. Wang, X.Y.; Zhu, Z.M.; Zhou, F.Q.; Zhang, F.M. Complete calibration of a structured light stripe vision sensor through a single cylindrical target. *Opt. Lasers Eng.* **2020**, *131*, 106096. [CrossRef]
37. Li, L.; Lin, B.Q.; Zou, Y.B. Study on Seam Tracking System Based on Stripe Type Laser Sensor and Welding Robot. *Chin. J. Lasers* **2015**, *42*, 34–41.
38. Kidong, L.; Insung, H.; Young-Min, K.; Huijun, L.; Munjin, K.; Jiyoung, Y. Real-Time Weld Quality Prediction Using a Laser Vision Sensor in a Lap Fillet Joint during Gas Metal Arc Welding. *Sensors* **2020**, *20*, 1625–1641.
39. Jawad, M.; Halis, A.; Essam, A. Welding seam profiling techniques based on active vision sensing for intelligent robotic welding. *Int. J. Adv. Manuf. Technol.* **2017**, *88*, 127–145.
40. Qiao, G.F.; Sun, D.L.; Song, G.M. A Rapid Coordinate Transformation Method for Serial Robot Calibration System. *J. Mech. Eng.* **2020**, *56*, 1–8.
41. Lei, T.; Huang, Y.; Wang, H.; Rong, Y.M. Automatic weld seam tracking of tube-to-tube sheet TIG welding robot with multiple sensors. *J. Manuf. Process.* **2020**, *3*, 47–52.
42. Kovacevic, R.; Zhang, S.B.; Zhang, M.Y. Noncontact Ultrasonic Sensing for Seam Tracking in Arc Welding Processes. *J. Manuf. Sci. Eng.* **1998**, *120*, 600–608.

Article

Sensor-Enabled Multi-Robot System for Automated Welding and In-Process Ultrasonic NDE

Momchil Vasilev, Charles N. MacLeod, Charalampos Loukas, Yashar Javadi *, Randika K. W. Vithanage, David Lines, Ehsan Mohseni, Stephen Gareth Pierce and Anthony Gachagan

Centre for Ultrasonic Engineering (CUE), Department of Electronic & Electrical Engineering, University of Strathclyde, Glasgow G1 1XQ, UK; momchil.vasilev@strath.ac.uk (M.V.); charles.macleod@strath.ac.uk (C.N.M.); charalampos.loukas@strath.ac.uk (C.L.); randika.vithanage@strath.ac.uk (R.K.W.V.); david.lines@strath.ac.uk (D.L.); ehsan.mohseni@strath.ac.uk (E.M.); s.g.pierce@strath.ac.uk (S.G.P.); a.gachagan@strath.ac.uk (A.G.)
* Correspondence: yashar.javadi@strath.ac.uk

Abstract: The growth of the automated welding sector and emerging technological requirements of Industry 4.0 have driven demand and research into intelligent sensor-enabled robotic systems. The higher production rates of automated welding have increased the need for fast, robotically deployed Non-Destructive Evaluation (NDE), replacing current time-consuming manually deployed inspection. This paper presents the development and deployment of a novel multi-robot system for automated welding and in-process NDE. Full external positional control is achieved in real time allowing for on-the-fly motion correction, based on multi-sensory input. The inspection capabilities of the system are demonstrated at three different stages of the manufacturing process: after all welding passes are complete; between individual welding passes; and during live-arc welding deposition. The specific advantages and challenges of each approach are outlined, and the defect detection capability is demonstrated through inspection of artificially induced defects. The developed system offers an early defect detection opportunity compared to current inspection methods, drastically reducing the delay between defect formation and discovery. This approach would enable in-process weld repair, leading to higher production efficiency, reduced rework rates and lower production costs.

Keywords: non-destructive evaluation; robotic NDE; robotic welding; robotic control; in-process NDE; ultrasonic NDE; ultrasound

Citation: Vasilev, M.; MacLeod, C.N.; Loukas, C.; Javadi, Y.; Vithanage, R.K.W.; Lines, D.; Mohseni, E.; Pierce, S.G.; Gachagan, A. Sensor-Enabled Multi-Robot System for Automated Welding and In-Process Ultrasonic NDE. *Sensors* **2021**, *21*, 5077. https://doi.org/10.3390/s21155077

Academic Editor: Salvatore Salamone

Received: 7 June 2021
Accepted: 21 July 2021
Published: 27 July 2021

Publisher's Note: MDPI stays neutral with regard to jurisdictional claims in published maps and institutional affiliations.

Copyright: © 2021 by the authors. Licensee MDPI, Basel, Switzerland. This article is an open access article distributed under the terms and conditions of the Creative Commons Attribution (CC BY) license (https://creativecommons.org/licenses/by/4.0/).

1. Introduction

The automated welding industry has been valued at USD 5.5 billion in 2018 and is expected to double by 2026, reaching USD 10.8 billion [1] with industrial articulated robots predicted to replace current traditional column and boom systems and manual operations. This growth has been driven by key high-value manufacturing sectors including automotive, marine, nuclear, petrochemical and defence. Paired with the technological demands of Industry 4.0 [2], the need for the development of intelligent and flexible sensor-enabled robotic welding systems has become paramount.

The wide adoption of automated manufacturing systems has subsequently raised the demand for automatically deployed and adaptive Non-Destructive Evaluation (NDE) in order to keep up with the faster production lines, when compared to manual manufacturing processes [3]. Developments in automated NDE are driven by industrial demand for fast and reliable quality control in high-value and high-throughput applications. In general, automatic systems provide greater positional accuracy, repeatability and inspection rates when compared to human operators, therefore, resulting in faster inspection speeds and reduced manufacturing costs. The ever-improving capabilities of such systems, on the other hand, lead to an overall increase in asset integrity and lifecycle, resulting in further long-term savings. Safety is another key advantage of automated NDE systems, as they can be deployed in hazardous environments, dangerous conditions and sites where human access is limited or not possible [4,5], thus improving working conditions and reducing the risks of workplace injuries and harmful substance exposure [6].

Single-axis scanners offer the ability for axial or circumferential scans of pipes and are suitable for on-site inspection of assets such as oil and gas pipelines. Such scanners can be guided by a track, or can be free-rolling where a projected laser line is used by the operator to positionally align the scanner with the weld [7,8]. Mobile crawler systems offer a higher degree of positional flexibility through a two-axis differential drive and can magnetically attach to the surfaces of assets enabling vertical deployment [9]. In addition, their compact size makes them well suited for remote applications with constrained access [4]. One particular challenge with such crawlers is accurately tracking their position, which is achieved through a combination of drive encoders, accelerometers, machine vision and in often cases expensive external measurement systems [10]. Multirotor aerial vehicles can deliver visual [11], laser and, more recently, contact ultrasonic [12] sensors in remote NDE inspection scenarios, where a magnetic crawler could not be deployed. While umbilical/tether cables are used commonly with mobile crawlers, they pose a challenge for the manoeuvrability and range of aerial systems. As a result, the power source, driving electronics and data storage for NDE sensors need to be on board the multirotor and, therefore, must be designed according to its limited payload capabilities. These systems can typically position and orient sensors in four axes (X, Y, Z and yaw) with recently developed over-actuated UAVs aiming to overcome this in support of omnidirectional contact-based airborne inspection [13].

Fixed inspection systems offer a higher degree of positional accuracy, compared to mobile systems. Gantries and cartesian scanners operate in a planar or boxed work envelope and are suited for components with simple geometries. Articulated robotic arms, on the other hand, operate in a spherical work envelope and enable the precise delivery of sensors in six Degrees of Freedom (DoF) with pose repeatability of under ±0.05 mm and maximum linear velocities of 2 m/s [14]. They are widely used in industry thanks to their flexibility and reprogrammability, and their positional repeatability makes them suited for operations with well controlled conditions such as component dimensions, position and orientation. Seven DoF robots are also available, with the additional seventh axis in the form of a linear track or a rotational joint allowing a wider range of robot poses to reach the same end-effector position, enabling the inspection of more complex structures.

As specified in the international standards for ultrasonic NDE of welds [15–17], joints of metals with a thickness of 8 mm or above are to be tested with shear waves, inserted through contact angled wedges, where the induced ultrasonic beam must have a normal angle of incidence with the weld interface. The ultrasonic probe must be moved across the surface of the sample in a way that provides full coverage of the weld joint. Alternatively, a sweep of multiple beams across a range of angles can be induced via beamforming through a Phased Array Ultrasonic Transducer (PAUT) [18], forming a sectorial scan. Moreover, PAUT probes enable the acquisition of all transmit–receive pairs through Full Matrix Capture (FMC), which offers the advantage of retrospective beamforming and reconstruction of the weld area through the Total Focusing Method (TFM) [19,20].

NDE is a particular bottleneck when considering high-value automated welding, as it is traditionally performed days after manufacturing when the parts are allowed to cool down [15,16], to ensure cooling-related defects are found. As such, any defects that are detected in the welds and do not pass an acceptance criteria [17] would either require the part to be sent back for repairs or, in some cases, would lead to scrapping the component altogether. Apart from adding to the overall production process inefficiency, this problem also results in higher production costs and longer, less consistent lead times. This, paired with the fact that welds of thicker components, large bore pipes and Wire + Arc Additive Manufacture (WAAM) parts [21] require days and, in some cases, weeks to complete, increases the need for fast in-process NDE inspection. By integrating the inspection into the manufacturing process, an early indication of potential defects can be obtained, effectively addressing the production and cost inefficiencies by allowing for defects to be qualified and potentially repaired in-process.

Current state-of-the-art robotic NDE systems and automated welding systems rely on robot controllers for calculating the kinematics and executing the motion, which are usually programmed by users manually jogging the robot to individual positions through a teaching pendant. Furthermore, emerging sensors, such as optical laser profiles and cameras can be utilised and deployed to provide real-time path correction. However, the deployment of application-specific sensors is highly dependent on the commercially available software provided by industrial robot manufacturers and the supported communication protocols. Therefore, it would be particularly beneficial to bypass the internal motion planning of a robotic controller and to apply external real-time positional control, based on additional sensor inputs, effectively shifting the path planning and sensor integration to another controller. In particular, the Robot Sensor Interface (RSI) [22] communication protocol could be leveraged in order to provide such an external positional control capability.

RSI was developed by industrial robot manufacturer KUKA for influencing a pre-programmed motion path through sensor input in order to achieve an adaptive robotic behaviour. The protocol is based on an interpolation cycle, which executes in real-time intervals of 4 ms for KRC (KUKA Robot Controller) 4 controller-based robots, and 12 ms for legacy KRC 2-based robots. During this, an XML string with a special format is transmitted over a UDP (User Datagram Protocol) link between the robotic controller and an external sensor or system. In [3], RSI was used in conjunction with a force-torque sensor to maintain constant contact force between a composite wing component and an ultrasonic roller probe, effectively accounting for any discrepancies between the CAD model of the part and the as-built geometry. This method, however, required that the motion path is pre-set within a robotic program, making use of the built-in KUKA trajectory planning algorithm. In [23], a custom trajectory planning algorithm was developed and embedded on a KRC 4 controller through a real-time RSI configuration diagram. This gave the capability to dynamically set and update the target position over Ethernet and the layer of abstraction based on a C++ Dynamic Link Library (DLL), made it possible to utilise the toolbox in various programming environments, e.g., MATLAB, Python and LabVIEW. Although providing a fast response time, the toolbox did not have a provision for real-time motion correction based on sensory input and was fully reliant on the KRC for execution.

This paper presents the development of a sensor-enabled multi-robot system for automated welding and in-process ultrasonic NDE. Table 1 shows a comparison between this work and state-of-the-art commercial robotic NDE systems, i.e., Genesis Systems NSpect [24], TWI IntACom [25], Tecnatom RABIT [26], FRS Robotics URQC [27] and Spirit AeroSystems VIEWS [3]. A novel sensor-driven adaptive motion algorithm for the control of industrial robots has been developed. Full external positional control was achieved in real time allowing for on-the-fly motion correction, based on multi-sensory input. A novel multi-robot welding and NDE system was developed, allowing for the flexible manufacture of welded components and the research into, and deployment of, NDE techniques at the point of manufacture. Thus, the automatic high-temperature PAUT inspection of multi-pass welded samples at three distinct points of the welding manufacture has been made possible, for the first time: inspection of the hot as-welded components; interpass inspection, between welding pass deposition; and live-arc inspection, in parallel with the weld deposition. Through the insertion of artificially induced defects, it has been demonstrated that in-process ultrasonic inspection is capable of early defect detection, drastically reducing the delay between defect formation and discovery. Furthermore, the developed system has enabled the real-time control of the welding process through live-arc ultrasonic methods. Conventional PAUT and FMC are made possible through a high-speed ultrasonic phased array controller, allowing for the use of advanced image processing algorithms, producing results which cannot be achieved using conventional ultrasonics. The work presented herein has directly supported and enabled further research into in-process weld inspection, across sectors, with the aim of producing right-first-time welds. As a result, it is envisaged that future High Value Manufacturing (HVM) of welded components will have an increased component quality, process efficiency, and reduced rework rates, lead-time inconsistencies and overall costs.

Table 1. Comparison between state-of-the-art commercial robotic NDE systems and this work.

	NSpect	IntACom	RABIT	URQC	VIEWS	This Work
Automated robotic NDE	✓	✓	✓	✓	✓	✓
Adaptive motion	✗	✗	✓	✓	✓	✓
FMC capture	✗	✓	✗	✗	✓	✓
Real-time trajectory control	✗	✗	✗	✗	✗	✓
Sensor integration independent from robot controller	✗	✗	✗	✗	✗	✓
NDE integrated with manufacture	✗	✗	✗	✗	✗	✓
High temperature inspection	✗	✗	✗	✗	✗	✓

Where ✓ denotes yes and ✗ denotes no.

2. Experimental System

2.1. Hardware

The automated welding and NDE system depicted in Figure 1 is based around a National Instruments cRIO 9038 [28] real-time embedded controller. The cRIO features a real-time processor and a Field-Programmable Gate Array (FPGA) on board, which enables fast, real-time parallel computations. Eight expansion slots for additional Input/Output modules enable direct sensor connectivity in addition to the Ethernet, USB and other interfaces, featured on the cRIO. The expansion modules used were the NI 9476 Digital Output, NI 9263 Analogue Output, NI 9205 Analogue Input, NI 9505 DC Motor Drive and an NI 9214 Thermocouple module.

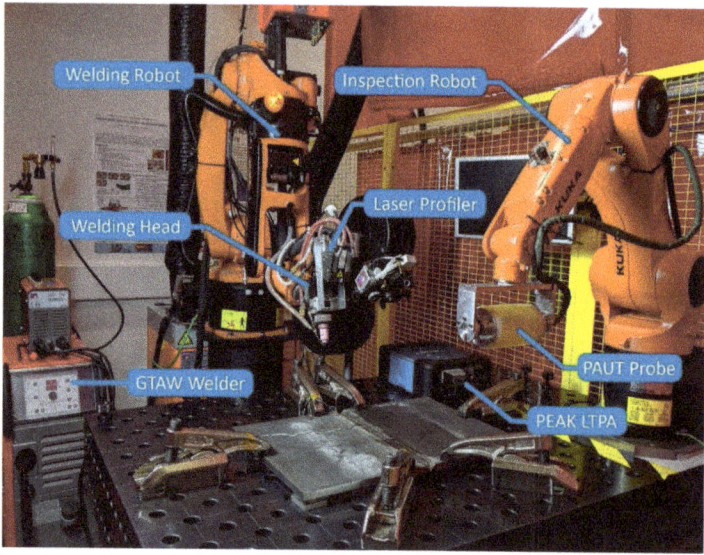

Figure 1. Sensor-enabled multi-robot welding and in-process NDE system.

Automation was implemented through two 6 DoF industrial manipulators, controlled in real time through RSI over an Ethernet connection. A KUKA KR5 Arc HW with a KRC 2 controller was employed as the Welding Robot (WR), while a KUKA AGILUS KR3 with a KRC 4 controller was employed as the Inspection Robot (IR). The welding hardware comprised of a JÄCKLE/TPS ProTIG 350A AC/DC [29] welding power source and a TBi Industries water-cooled welding torch, mounted on the welding robot end effector. The welding arc was triggered through a 24 V digital signal connected to the power source, while the arc current was set through a 10 V differential analogue line. The power source featured process feedback in the form of measured arc current and arc voltage, also transmitted through differential analogue lines. A JÄCKLE/TPS 4-roll wire feeder, with an optical encoder was powered and controlled via the NI 9505. Its rotational speed was measured and controlled using Pulse Width Modulation (PWM) and was related appropriately to the desired control metric of linear wire feed rate. A Micro-Epsilon scanCONTROL 9030 [30] laser profiler was utilised for weld seam tracking and measurement, while an XIRIS XVC 1100 [31] high dynamic range weld monitoring camera provided visual feedback of the process.

The workpiece temperature was measured through permanently attached thermocouples, which were used to maintain the workpiece within a desired interpass temperature range. The thermocouples were also utilised for monitoring the temperature gradient across the workpiece, which is a crucial requirement for temperature compensation of the ultrasonic images. A high-temperature PAUT roller probe was attached to the flange of the IR driven by a PEAK LTPA [32] low-noise ultrasonic phased array controller. The bandwidth and storage of the cRIO were only sufficient for inspection with conventional UT probes, therefore, the LTPA had to be directly connected to the host PC when using phased array probes. The bandwidth challenge could be addressed by substituting the cRIO with a high-performance NI PXI real-time controller. Finally, the Graphic User Interface (GUI) was deployed on the host PC, facilitating the user input, process monitoring and control. The high-level system architecture is shown in Figure 2, where the hardware components are represented by blue blocks, the software tasks are represented by green blocks and the communication links are shown as arrows.

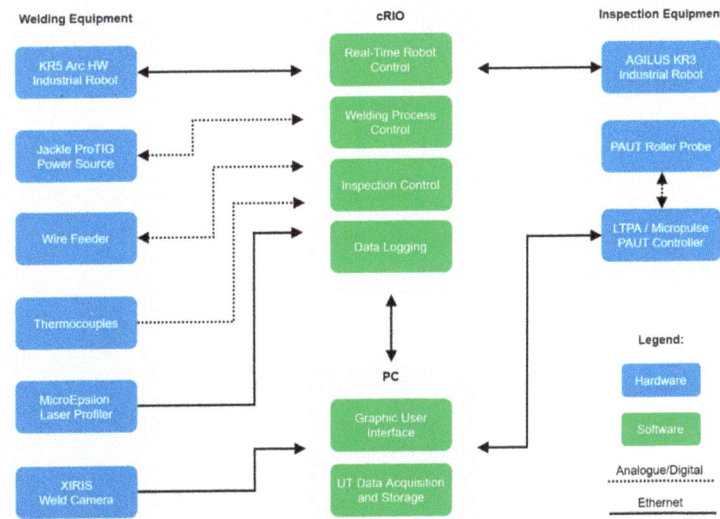

Figure 2. Sensor-enabled multi-robot welding and in-process NDE system architecture. Overall process control was implemented on the NI cRIO, while the GUI and PAUT acquisition and storage were executed on a host PC.

2.2. Software

All software was developed in the cRIO native LabVIEW environment which enabled rapid prototyping, due to the wide range of supported communication protocols and software libraries. The software architecture was built using the JKI state machine [33] and parallel real-time Timed Loops, ensuring program flexibility while also providing reliable and fast response times. Three parallel state machines were responsible for executing the program sequence, controlling the Welding Robot (WR) and controlling the Inspection Robot (IR), respectively.

2.2.1. Real-Time Robotic Control

The real-time robotic control strategy employed full external positional control of the robots. This was achieved through a correction-based RSI motion, meaning that the robot controller did not hold any pre-programmed path, and the robot end-effector position was updated on-the-fly through positional corrections. At every iteration of the interpolation cycle, the current position and timestamp of the internal clock are sent by the robot controller as an XML string. An XML string response is returned by the cRIO, mirroring the timestamp to keep the connection alive, and providing positional corrections in each axis, which determine where the end-effector will move to over the next interpolation cycle. There are two types of positional corrections—absolute, where the new position is given with respect of the robot base, and relative, where the new position is given with respect to the current position. For example, an absolute correction of 1 mm in the X-axis will move the end-effector to the absolute coordinate $X = 1$ mm, while the same relative correction will move the robotic end-effector by 1 mm in the positive X-axis direction irrespective of its current position. Relative corrections were chosen for this body of work as the smaller magnitude of corrections sent to the robot controller made them safer for use during the development and testing stage.

Welding and inspection robot paths are inputted by the user as individual points in a table through the GUI, where each row corresponds to a point in the path, while the columns hold the cartesian coordinates for each axis. Additional columns in the welding path table provide control over the process while approaching the target, i.e., an "Arc On" Boolean determines if the WR should be welding, and a "Log On" Boolean enables the data logging. More sophisticated data can also be included as additional columns, for example, to choose the welding parameters through a lookup table containing the settings for root, hot, filling and cap passes, therefore allowing the user to enter the parameters from a relevant Welding Procedure Specification (WPS) document alongside the robotic path. When considering simpler geometries such as a plate or pipe butt-weld, the robotic paths can be manually entered as individual point coordinates; for example, a straight-line weld would only require two points—the start and the end of the weld. For more complex geometries this can be generated by Computer Aided Manufacture (CAM) or robotic path planning software and imported into the software [34–36].

2.2.2. Trajectory Planning

An on-the-fly calculated trajectory planning algorithm running at the RSI interpolation cycle rate was implemented as demonstrated in Figure 3. A relative positional correction is sent to the KRC at each iteration of the interpolation cycle, consisting of a linear motion component d_L and an adaptive motion component d_A. The Linear Motion Controller (LMC) is responsible for executing a straight-line trajectory between the current end-effector position P_C and a target position $P_T{'}$. It is based on a linear acceleration–cruise–deceleration curve with the setpoint cruise speed V entered by the user. The Adaptive Motion Controller (AMC) generates an instantaneous adaptive correction d_A in response to the sensory input and process requirements. The absolute adaptive correction D_A, which is the cumulative total correction that has been applied by the AMC, is summed to the current target position P_T taken from the robot path table to form $P_T{'}$.

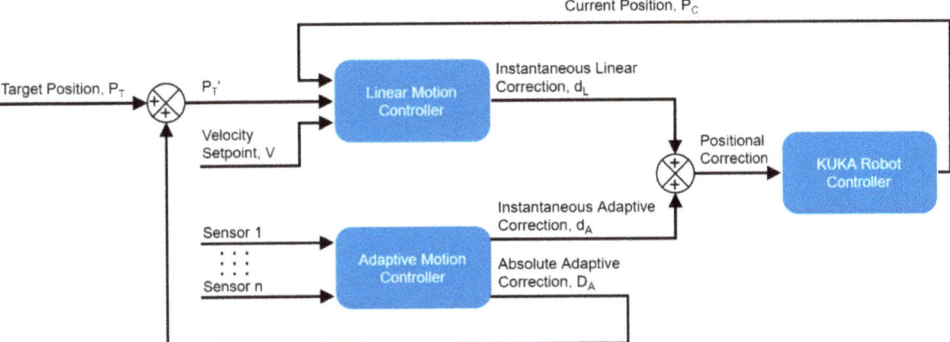

Figure 3. Trajectory planning and on-the-fly sensor-based motion correction algorithm.

Figure 4a shows the operation of the LMC with an example linear trajectory along the X-axis between a starting point P_S and a termination point P_T. The linear motion velocity vector V_L at an arbitrary point P_0 along the path is always directed towards the target point P_T and is therefore parallel and coinciding with the $P_S P_T$ vector. Furthermore, as the $P_S P_T$ vector is aligned with the X-axis in Figure 4, the V_L vector only consists of an X-axis component. In Figure 4b, an example AMC output d_A, consisting of a sinusoidal oscillation in the Y-axis, is summed with d_L before sending the positional correction to the KRC, resulting in a weaving motion between P_S and P_T. However, as the linear motion vector V_L is always directed towards the target P_T, a Y-axis component is introduced at all points that do not lie on the $P_S P_T$ vector, which results in a distorted trajectory. The effects of this distortion become stronger and more evident closer to P_T as illustrated by V_{L0} and V_{L1} in Figure 4b. In order to avoid the distortion in the LMC trajectory caused by the instantaneous correction d_A, the absolute adaptive correction D_A is summed with P_T to give P_T'. This offsetting of the target point ensures that the LMC-generated trajectory remains linear as shown in Figure 4c. As a result, a trade-off between target point accuracy and adaptive correction is inherently introduced in the system.

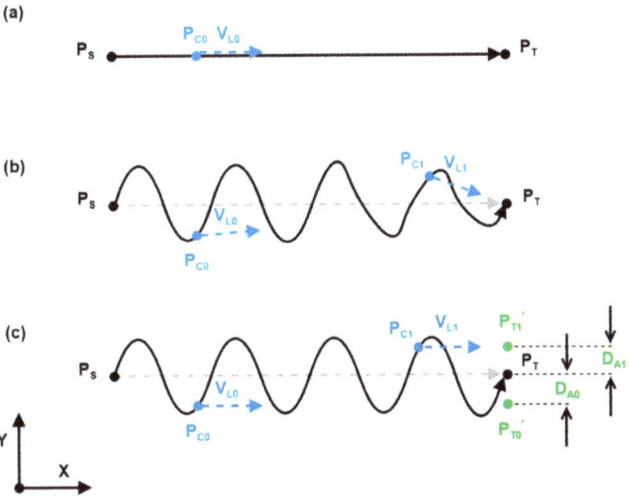

Figure 4. (a) Example linear motion generated by the LMC; (b) trajectory distortion introduced by instantaneous adaptive correction d_A; (c) target point offsetting through absolute adaptive correction D_A.

The demonstrated weaving motion is useful in various scenarios; for example, in welding, when mimicking the motion of manual welding techniques. Such a weaving motion is generally not achievable through a robotic teach pendant and requires path planning software. The software would normally create the path through a number of fundamental linear and circular motions, which would require a full trajectory recalculation if any of the parameters such as the travel speed, amplitude or frequency of weaving need to be modified. In contrast, as the weaving motion is calculated in real time, its parameters and driving function can be readily changed and updated on-the-fly. This approach can be applied to multiple axes at the same time and can be implemented with multiple sensors. For example, most modern automated welding power supplies offer the ability to monitor the arc current and arc voltage in real time, which can be utilised for process control. The measured arc voltage in the Gas Tungsten Arc Welding (GTAW) process is directly correlated to the distance between the welding torch and the workpiece, and as such is suitable for adaptive motion. When welding a workpiece that is assumed to be flat, but has surface height variations, the offset between the welding torch and the sample surface would vary along the weld as shown in Figure 5a, resulting in an inconsistent arc voltage and, therefore, inconsistent weld properties. The measured arc voltage was used as the control variable of a Proportional–Integral–Derivative (PID) control loop, the output of which was an instantaneous adaptive correction applied in the Z-axis. This allowed for Automatic Voltage Control (AVC), subsequently maintaining that the welding torch to workpiece distance is constant as illustrated in Figure 5b. The demonstrated approach can be applied for a variety of scenarios with equipment such as laser profilers, force-torque sensors and machine vision cameras among others.

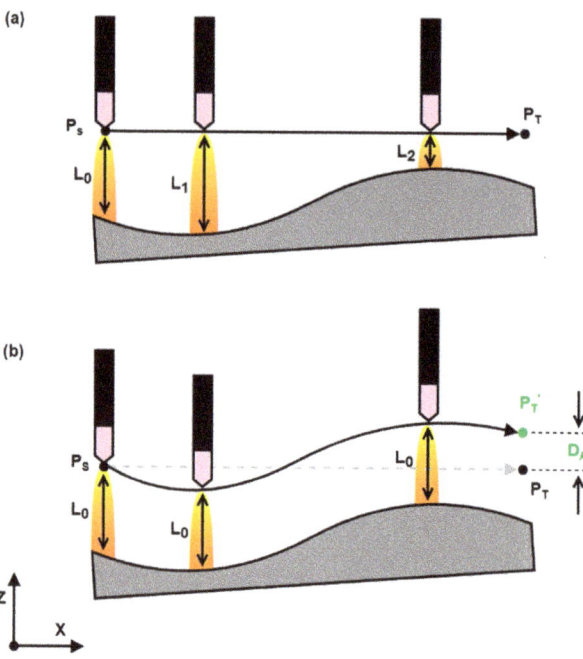

Figure 5. (a) Open-loop welding of a sample with an uneven surface through a linear trajectory; the welding torch to sample distance changes along the weld; (b) closed-loop welding of a sample with an uneven surface through an adaptive trajectory; on-the-fly adjustment of torch offset is achieved through the measured arc voltage; the welding torch to sample distance is constant along the weld; the end point P_T is shifted to P_T' as a result of the adaptive motion.

2.2.3. Welding Sequence

All relevant process parameters and ultrasonic measurements were timestamped, positionally encoded by the robot position and logged in a binary format for subsequent analysis. Before any welding, the WR performed a calibration using the laser profiler in order to measure and locate the weld groove. This was performed only once, as the workpieces were fixed to the table using 6-point clamping and their location was not expected to shift with respect to the WR. In applications where an initial scan of the weld groove is not practical, or where the weld groove is expected to shift, the welding system has the capability to utilise the laser profiler output for real-time seam tracking, through the AMC. All multipass welding and inspection trials were performed on 15 mm thick S275 structural steel plates, bevelled to form a 90° V-groove. The plates were butt-welded by the WR over a total of 21 passes deposited over 7 layers, as shown in Figure 6.

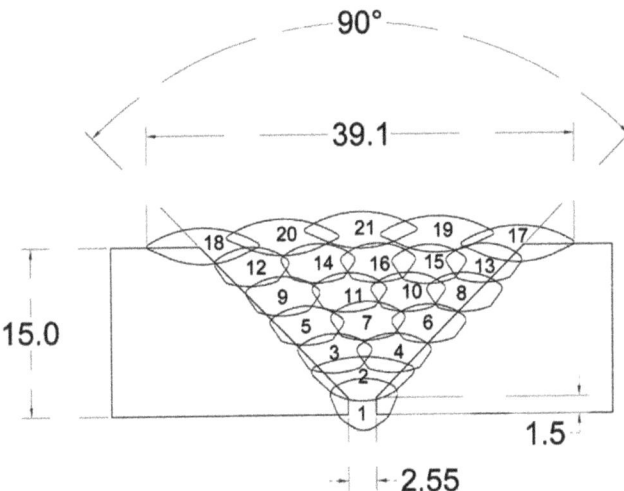

Figure 6. Multipass weld specification for 15 mm thick S275 steel bevelled with a 90° V-groove; a total of 21 passes are deposited over 7 layers; all linear dimensions are in millimetres.

3. Ultrasonic Inspection

The system was developed with the aim to perform ultrasonic inspections at three distinct points of the welding process: post-process, when all welding is completed; interpass in-process, between distinct welding passes; and live-arc in-process, in parallel with the weld deposition. Despite the distinct advantages and disadvantages of each approach, they would all fundamentally lead to early defect detection.

3.1. Post-Process UT

The accuracy and positional repeatability of robots can be leveraged for post-process NDE by performing continuous repeated inspections of the as-built component. This allows for the development of any defects such as cold cracking to be monitored by comparing successive ultrasonic images. Due to the elevated sample temperature introduced by the welding process and any post-heat treatment, a high-temperature capable ultrasonic probe assembly was necessary. An Olympus 5L64-32 × 10-A32-P-2.5-HY array (5 MHz, 64 element, 0.5 mm element pitch, 10 mm element elevation) was used in conjunction with an SA32C-ULT-N55S-IHC angled wedge (suited for shear wave inspection in steel centred around 55°). The wedge is manufactured out of the material ULTEM and so is capable of short-term contact temperatures of up to 150 °C. High-temperature ultrasonic couplant was used between the transducer and wedge. Before touching down on each inspection position, the ultrasonic wedge was dipped in a custom-designed tray containing the same high-temperature ultrasonic couplant to ensure good acoustic propagation between probe and sample. Figure 7 shows the detection and growth monitoring of a hydrogen crack that was artificially induced in the Heat Affected Zone (HAZ), adjacent to the weld toe, through localised water quenching [37].

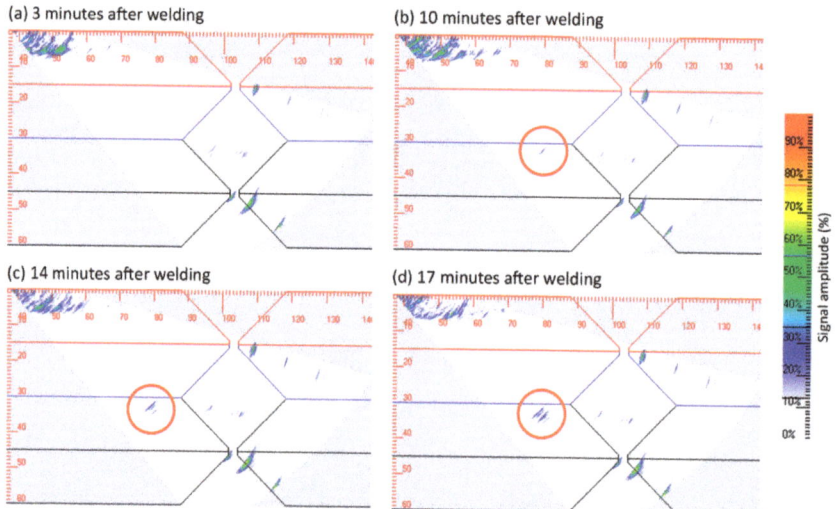

Figure 7. Continuous post-weld ultrasonic imaging of artificially induced hydrogen crack. The crack was initiated 10 min after all welding passes were deposited and its growth was observed in time. The location of the crack was in the HAZ adjacent to the weld cap toe.

The elevated temperature of the sample after it is manufactured must be taken into account when performing NDE as the speed of sound in the material varies with temperature. As the sample cools down, this causes imaging anomalies in both amplitude and position. In [38], a Tungsten rod was introduced in the weld to form a static reflector of known size and location [39]. The weld was repeatedly inspected at regular time intervals for a period of 22 h, and the position and amplitude of the inserted reflector were extracted to form a thermal compensation curve. The sample temperature at the inspection location decreased from 164 °C at 2 min after welding to 28 °C at 75 min after welding. As a result, the reflected amplitude increased significantly from 25% to 62% of full screen height, and the defect indication's position shifted by 3 mm on the reconstructed sector scan image. These data were utilised to correct the position and amplitude of an artificially induced crack. The crack initiation was successfully detected 22 min after the weld completion, and it was observed to be growing over a total of 90 min.

3.2. Interpass In-Process UT

Interpass ultrasonic NDE allows for the detection of weld flaws through inspection between individual welding passes or layers and provides an opportunity for in-process repair, as only a small amount of material would need to be removed in order to excavate and repair the defects. This is particularly advantageous for the manufacture of components that are typically challenging to repair after all welding passes have been deposited, e.g., thick multipass welds and WAAM parts. A key challenge of interpass welding inspection is the complex sample geometry which changes as the weld is deposited and therefore differs from the as-built geometry [40]. Figure 8 shows that the unwelded portion of the V-groove in a multipass weld causes a number of reflections and artefacts in the ultrasonic images, as demonstrated at three distinct stages of the sample manufacture. As the weld is deposited, the sample geometry reflections change in shape and size, until they disappear upon completion of the weld joint. Hence, appropriate signal processing and masking are required to effectively remove the false positive indications from the sample geometry.

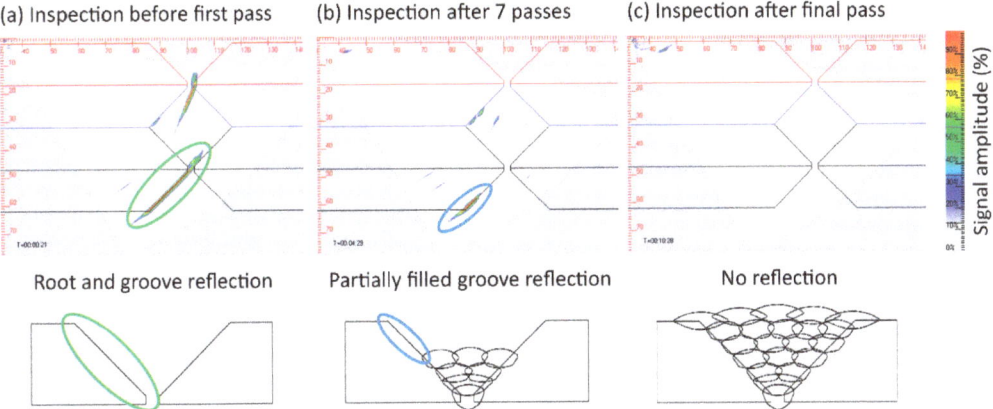

Figure 8. Ultrasonic sectorial scan of 90° V-groove multipass weld; (**a**) before welding the groove edge is detected as a reflector (green marker); (**b**) after 7 passes are deposited, the size of the groove edge indication is reduced (blue marker); (**c**) after all welding passes are deposited, the groove edge is no longer detected.

The high interpass temperatures required to maintain the weld integrity (typically up to 250 °C) have driven research into the development of a novel, high-temperature capable PAUT probe [41]. The probe features a 5 MHz, linear 64-element PAUT transducer immersed in water and enclosed in a moulded high-temperature silicone rubber tyre, capable of operating at temperatures up to 350 °C. Coupling between the probe and the sample was achieved through a constant compressional force and high-temperature gel couplant as demonstrated in Figure 9. The novel probe has allowed for the interpass detection of artificially induced defects inside a partially filled multipass weld such as the one shown in Figure 10, where a Tungsten rod with a diameter of 2.4 mm and length of 30 mm was included in the weld.

Figure 9. Interpass in-process UT inspection with a novel high-temperature PAUT roller probe.

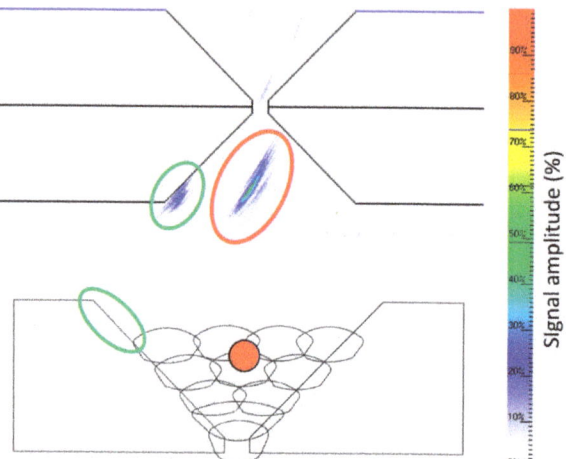

Figure 10. Interpass ultrasonic image of artificially induced defect (Tungsten rod with 2.4 mm diameter) (red marker) with a false positive indication from the unwelded groove edge (green marker).

As a result of the moving heat source in welding, thermal gradients in both the direction of welding and perpendicular to the direction of travel are introduced in the workpiece, ultimately resulting in ultrasonic image distortion. Furthermore, the dynamic nature of multipass welding essentially results in a different thermal gradient after each welding pass. An in-process thermal compensation procedure was proposed in [42] involving the parallel manufacture of a second, identical sample with an embedded Tungsten pipe, serving as an in-process calibration block. The reflection from the known in size and location pipe was used to calibrate for the effects of the temperature gradients and it was demonstrated that the approach provided more accurate results, compared to a traditional calibration on a sample with a side drilled hole at a uniform temperature. For the most accurate calibration and thermal compensation results, however, the sample temperature would need to be precisely known through a combination of measurement and weld modelling [43]. It is important to also note that interpass inspection could increase the component manufacture duration, as it is deployed sequentially with the welding deposition. In addition, increasing the interval between welding passes could lead to excessive sample cooling and the loss of interpass temperature. Therefore, the UT acquisition and image processing speed must be taken into account when considering the deployment of interpass NDE for welding applications.

3.3. Live-Arc In-Process UT

In-process UT deployed during the welding deposition offers rapid feedback for the welding process and allows not only measurement, but also control of the welding process. In [44], a pair of air-coupled ultrasonic transducers were used to induce guided Lamb waves through a section of 3 mm thick plate butt joint while it was deposited. Figure 11 shows that the solidification of the weld was monitored in real time through live-arc in-process UT. This method has shown promise as the rate of change of the received Lamb waves' amplitude was found to be correlated to the weld penetration depth. In [45], a split-crystal ultrasonic wheel probe was attached to the welding torch and was utilised for thickness measurement of samples with a varying loss of wall thickness, as shown in Figure 12. The measured thickness was used to control and adapt the welding arc current, torch travel speed and wire feed rate on-the-fly. It was demonstrated that the approach provided sufficiently low latency and high accuracy for real-time welding process control and, as a result, provided a better performance of welding samples with thickness variations, compared to a traditional open-loop automated welding system.

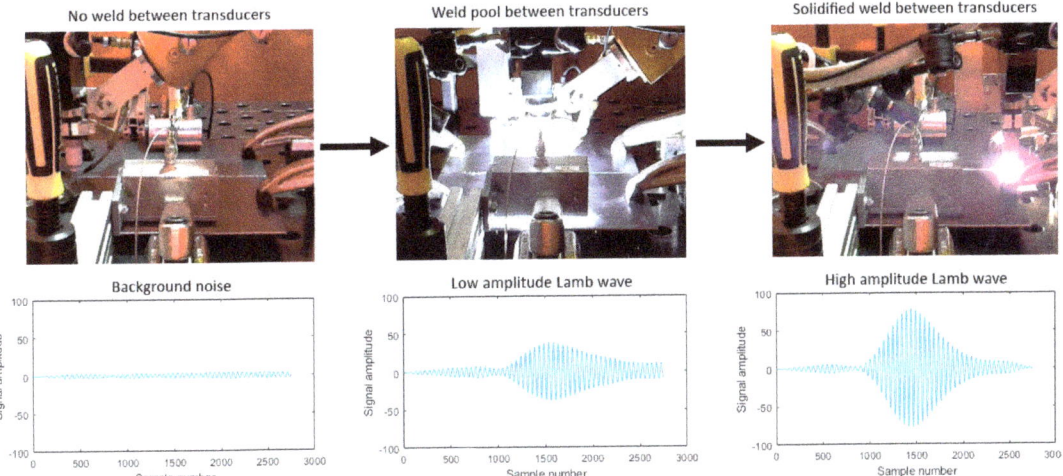

Figure 11. Live-arc in-process weld UT with non-contact Lamb waves.

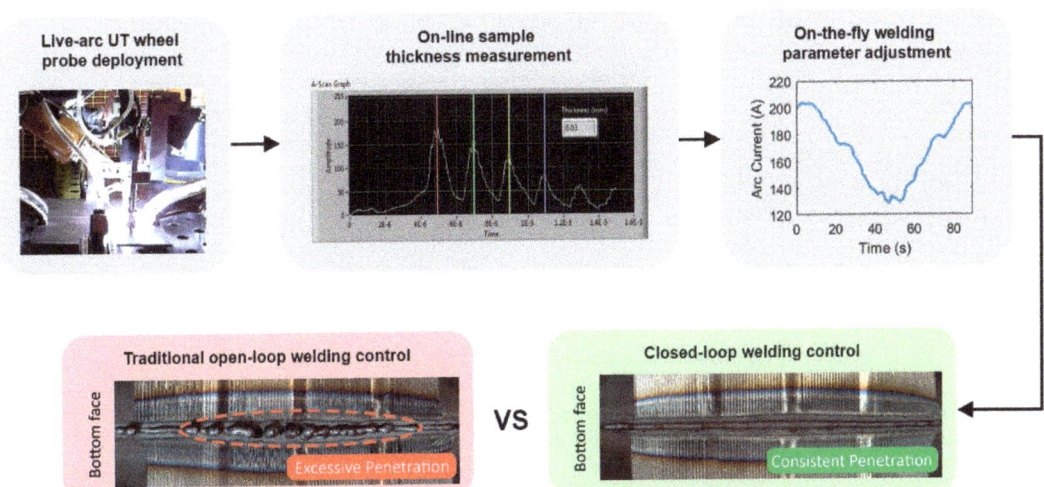

Figure 12. On-the-fly adaptive welding control through live-arc UT sample measurement.

Current work in the University of Strathclyde is focused on addressing the challenges associated with deploying PAUT probes during the weld deposition (Figure 13). The next generation of PAUT probes will be dry coupled, which would remove the risk of unwanted weld contamination by the ultrasonic gel that can cause porosity [37] and would reduce the variation in coupling between the probe and the workpiece.

Figure 13. Live-arc PAUT inspection experiment.

4. Conclusions and Future Work

A novel sensor-enabled robotic system for automated welding and ultrasonic inspection was developed and evaluated. The system architecture was based around an NI cRIO real-time embedded controller which enabled real-time communication, data acquisition and control. A real-time external robotic control strategy for adaptive behaviour was developed, allowing for on-the-fly sensor-based trajectory corrections. The inspection capabilities of the system were demonstrated in three different scenarios:

1. Post-process continuous UT—the initiation and growth of cold crack defects was observed and measured through a continuous inspection at regular intervals after the multipass weld was completely filled.
2. Interpass in-process UT—the challenges due to the complex sample geometry of the unfilled weld groove were demonstrated and the inspection results showed the defect detection capabilities through artificially induced defects.
3. Live-arc in-process UT—the deployment and application of three different ultrasonic sensors during live-arc welding deposition was outlined and the challenges and results were discussed.

Current work on masking the bevel edge reflections will remove false positives arising from the unfilled weld groove and thermal gradient compensation would enable the accurate locating and sizing of weld defects. Future developments of the PAUT probe will allow for a completely dry coupled inspection, eliminating the coupling and contamination challenges posed by the ultrasonic couplant. It is envisaged that future welding and live-arc in-process systems would possess the capability for automatic in-process defect detection, which would in turn significantly reduce the delay between the development and detection of a defect, offering the potential for in-process weld repair.

Author Contributions: Conceptualisation, M.V., C.N.M., Y.J., R.K.W.V., D.L., E.M., S.G.P. and A.G.; software, M.V., C.L. and D.L.; investigation, M.V., C.L., Y.J., R.K.W.V., D.L. and E.M.; resources, C.N.M., S.G.P. and A.G.; writing—original draft preparation, M.V.; writing—review and editing, C.N.M., C.L., D.L. and E.M. All authors have read and agreed to the published version of the manuscript.

Funding: This work was funded by the Engineering and Physical Sciences Research Council (EPSRC), grant number 2096856.

Institutional Review Board Statement: Not applicable.

Informed Consent Statement: Not applicable.

Data Availability Statement: Not applicable.

Conflicts of Interest: The authors declare no conflict of interest.

References

1. Chinchane, A.; Onkar, S. Robotic Welding Market Size, Share I Welding Robot Statistics by 2026. Available online: https://www.alliedmarketresearch.com/robotic-welding-market (accessed on 17 March 2021).
2. What is Industry 4.0? How Does it Work? (A Beginners Guide). Available online: https://www.twi-global.com/what-we-do/research-and-technology/technologies/industry-4-0.aspx (accessed on 8 April 2021).
3. Mineo, C.; MacLeod, C.; Morozov, M.; Pierce, S.G.; Lardner, T.; Summan, R.; Powell, J.; McCubbin, P.; McCubbin, C.; Munro, G.; et al. Fast ultrasonic phased array inspection of complex geometries delivered through robotic manipulators and high speed data acquisition instrumentation. In Proceedings of the 2016 IEEE International Ultrasonics Symposium (IUS), Tours, France, 18–21 September 2016; pp. 1–4.
4. Dobie, G.; Summan, R.; Pierce, S.G.; Galbraith, W.; Hayward, G. A Noncontact Ultrasonic Platform for Structural Inspection. *IEEE Sens. J.* **2011**, *11*, 2458–2468. [CrossRef]
5. Hashem, J.A.; Pryor, M.; Landsberger, S.; Hunter, J.; Janecky, D.R. Automating High-Precision X-Ray and Neutron Imaging Applications With Robotics. *IEEE Trans. Autom. Sci. Eng.* **2018**, *15*, 663–674. [CrossRef]
6. Stern, R.M.; Berlin, A.; Fletcher, A.; Hemminki, K.; Jarvisalo, J.; Peto, J. International conference on health hazards and biological effects of welding fumes and gases. *Int. Arch. Occup. Environ. Heath* **1986**, *57*, 237–246. [CrossRef]
7. OLYMPUS WeldROVER Scanner. Available online: https://www.olympus-ims.com/en/scanners/weldrover/ (accessed on 13 May 2021).
8. JIREH Industries NAVIC—Weld Scanner. Available online: //www.jireh.com/products/navic-weld-scanner/ (accessed on 17 May 2021).
9. Eddify Scorpion 2 Ultrasonic Tank Shell Inspection I UT Thickness Readings. Available online: https://www.eddyfi.com/en/product/scorpion-2 (accessed on 17 May 2021).
10. McGregor, A.; Dobie, G.; Pearson, N.R.; MacLeod, C.N.; Gachagan, A. Determining Position and Orientation of a 3-Wheel Robot on a Pipe Using an Accelerometer. *IEEE Sens. J.* **2020**, *20*, 5061–5071. [CrossRef]
11. Zhang, D.; Watson, R.; Dobie, G.; MacLeod, C.; Khan, A.; Pierce, G. Quantifying impacts on remote photogrammetric inspection using unmanned aerial vehicles. *Eng. Struct.* **2020**, *209*, 109940. [CrossRef]
12. Zhang, D.; Watson, R.; MacLeod, C.; Dobie, G.; Galbraith, W.; Pierce, G. Implementation and evaluation of an autonomous airborne ultrasound inspection system. *Nondestruct. Test. Eval.* **2021**, 1–21. [CrossRef]
13. Watson, R.J.; Pierce, S.G.; Kamel, M.; Zhang, D.; MacLeod, C.N.; Dobie, G.; Bolton, G.; Dawood, T.; Nieto, J. Deployment of Contact-Based Ultrasonic Thickness Measurements Using Over-Actuated UAVs. In *Proceedings of the European Workshop on Structural Health Monitoring*; Rizzo, P., Milazzo, A., Eds.; Springer International Publishing: Cham, Switzerland, 2021; pp. 683–694.
14. KUKA AG Industrial Robotics_Low Payloads. Available online: https://www.kuka.com/-/media/kuka-downloads/imported/9cb8e311bfd744b4b0eab25ca883f6d3/kuka_robotics_low_payloads.pdf?rev=cbf117123ca142dda4c7abe9ba0a3e64&hash=2EC50752C478393CC7DD0E76F272BC59 (accessed on 19 May 2021).
15. Non-Destructive Testing of Welds—Ultrasonic Testing—Techniques, Testing Levels, and Assessment (BS EN ISO 17640-2018). Available online: https://shop.bsigroup.com/ProductDetail?pid=000000000030376825 (accessed on 22 July 2021).
16. Non-Destructive Testing of Welds—Ultrasonic Testing—Use of Automated Phased Array Technology (BS EN ISO 13588:2019). Available online: https://shop.bsigroup.com/ProductDetail?pid=000000000030353054 (accessed on 22 July 2021).
17. Non-Destructive Testing of Welds—Phased Array Ultrasonic Testing (PAUT)—Acceptance Levels (BS EN ISO 19285:2017). Available online: https://shop.bsigroup.com/ProductDetail?pid=000000000030342680 (accessed on 22 July 2021).
18. Drinkwater, B.W.; Wilcox, P.D. Ultrasonic arrays for non-destructive evaluation: A review. *NDT E Int.* **2006**, *39*, 525–541. [CrossRef]
19. Holmes, C.; Drinkwater, B.W.; Wilcox, P.D. Post-processing of the full matrix of ultrasonic transmit–receive array data for non-destructive evaluation. *NDT E Int.* **2005**, *38*, 701–711. [CrossRef]
20. Wilcox, P.D.; Holmes, C.; Drinkwater, B.W. Enhanced Defect Detection and Characterisation by Signal Processing of Ultrasonic Array Data. In Proceedings of the 9th European Conference on NDT, Berlin, Germany, 25–29 September 2006.
21. Williams, S.W.; Martina, F.; Addison, A.C.; Ding, J.; Pardal, G.; Colegrove, P. Wire + Arc Additive Manufacturing. *Mater. Sci. Technol.* **2016**, *32*, 641–647. [CrossRef]
22. KUKA.RobotSensorInterface 4.0. 2018. Available online: https://www.kuka.com/en-us/products/robotics-systems/software/system-software/kuka_systemsoftware (accessed on 22 July 2021).
23. Mineo, C.; Vasilev, M.; Cowan, B.; MacLeod, C.N.; Pierce, S.G.; Wong, C.; Yang, E.; Fuentes, R.; Cross, E.J. Enabling robotic adaptive behaviour capabilities for new Industry 4.0 automated quality inspection paradigms. *Insight* **2020**, *62*, 338–344. [CrossRef]
24. NSpect Systems I Robotic Non-Destructive Inspection Systems I Genesis. Genesis Systems. Available online: https://www.genesis-systems.com/robotic-integration/nspect-systems (accessed on 2 July 2021).
25. Mineo, C.; Pierce, S.G.; Wright, B.; Cooper, I.; Nicholson, P.I. PAUT inspection of complex-shaped composite materials through six DOFs robotic manipulators. *Insight Non Destr. Test. Cond. Monit.* **2015**, *57*, 161–166. [CrossRef]

26. Robot-based solution To Obtain an Automated, Integrated And Industrial Non-Destructive Inspection Process. Available online: https://www.ndt.net/search/docs.php3?id=16960&msgID=0&rootID=0 (accessed on 2 July 2021).
27. FRS Robotics Ultrasonic Robotic QC. Available online: https://www.frsrobotics.com/index.php/solutions/ultrasonic-robotic-qc (accessed on 2 July 2021).
28. National Instruments cRIO-9038. Available online: http://www.ni.com/en-gb/support/model.crio-9038.html (accessed on 5 September 2019).
29. Jäckle Schweiß- und Schneidtechnik GmbH Jackle ProTIG 350AC/DC Operating Manual. Available online: https://www.jess-welding.com/en/portfolio/protig-350-500/ (accessed on 9 May 2019).
30. Micro-Epsilon Compact Laser Scanner for High Precision. Available online: https://www.micro-epsilon.co.uk/2D_3D/laser-scanner/scanCONTROL-2900/ (accessed on 5 May 2020).
31. Xiris Automation Inc. XVC-1000/1100 Weld Camera. Available online: https://www.xiris.com/xiris-xvc-1000/ (accessed on 19 May 2021).
32. PeakNDT PEAK LTPA Specification. Available online: https://www.peakndt.com/products/ltpa/ (accessed on 5 September 2019).
33. JKISoftware/JKI-State-Machine 2021. Available online: https://github.com/JKISoftware/JKI-State-Machine (accessed on 8 April 2021).
34. Morozov, M.; Pierce, S.G.; MacLeod, C.N.; Mineo, C.; Summan, R. Off-line scan path planning for robotic NDT. *Measurement* **2018**, *122*, 284–290. [CrossRef]
35. Macleod, C.N.; Dobie, G.; Pierce, S.G.; Summan, R.; Morozov, M. Machining-Based Coverage Path Planning for Automated Structural Inspection. *IEEE Trans. Autom. Sci. Eng.* **2018**, *15*, 202–213. [CrossRef]
36. Mineo, C.; Pierce, S.G.; Nicholson, P.I.; Cooper, I. Robotic path planning for non-destructive testing—A custom MATLAB toolbox approach. *Robot. Comput. Integr. Manuf.* **2016**, *37*, 1–12. [CrossRef]
37. Javadi, Y.; Mohseni, E.; MacLeod, C.N.; Lines, D.; Vasilev, M.; Mineo, C.; Pierce, S.G.; Gachagan, A. High-temperature in-process inspection followed by 96-h robotic inspection of intentionally manufactured hydrogen crack in multi-pass robotic welding. *Int. J. Press. Vessel. Pip.* **2021**, *189*, 104288. [CrossRef]
38. Javadi, Y.; Mohseni, E.; MacLeod, C.N.; Lines, D.; Vasilev, M.; Mineo, C.; Foster, E.; Pierce, S.G.; Gachagan, A. Continuous monitoring of an intentionally-manufactured crack using an automated welding and in-process inspection system. *Mater. Des.* **2020**, *191*, 108655. [CrossRef]
39. Javadi, Y.; Vasilev, M.; MacLeod, C.N.; Pierce, S.G.; Su, R.; Mineo, C.; Dziewierz, J.; Gachagan, A. Intentional weld defect process: From manufacturing by robotic welding machine to inspection using TFM phased array. In Proceedings of the 45th Annual Review of Progress in Quantitative Nondestructive Evaluation, Burlington, VT, USA, 15–19 July 2018; p. 040011.
40. Lines, D.; Javadi, Y.; Mohseni, E.; Vasilev, M.; MacLeod, C.N.; Mineo, C.; Vithanage, R.K.W.; Qiu, Z.; Zimermann, R.; Loukas, C.; et al. A flexible robotic cell for in-process inspection of multi-pass welds. *Insight J. Br. Inst. Non Destr. Test.* **2020**, *62*, 526–532. [CrossRef]
41. Vithanage, R.K.W.; Mohseni, E.; Qiu, Z.; MacLeod, C.; Javadi, Y.; Sweeney, N.; Pierce, G.; Gachagan, A. A phased array ultrasound roller probe for automated in-process/interpass inspection of multipass welds. *IEEE Trans. Ind. Electron.* **2020**. [CrossRef]
42. Javadi, Y.; Sweeney, N.E.; Mohseni, E.; MacLeod, C.N.; Lines, D.; Vasilev, M.; Qiu, Z.; Vithanage, R.K.W.; Mineo, C.; Stratoudaki, T.; et al. In-process calibration of a non-destructive testing system used for in-process inspection of multi-pass welding. *Mater. Des.* **2020**, *195*, 108981. [CrossRef]
43. Mohseni, E.; Javadi, Y.; Sweeney, N.E.; Lines, D.; MacLeod, C.N.; Vithanage, R.K.W.; Qiu, Z.; Vasilev, M.; Mineo, C.; Lukacs, P.; et al. Model-assisted ultrasonic calibration using intentionally embedded defects for in-process weld inspection. *Mater. Des.* **2021**, *198*, 109330. [CrossRef]
44. Vasilev, M.; MacLeod, C.; Galbraith, W.; Javadi, Y.; Foster, E.; Dobie, G.; Pierce, G.; Gachagan, A. Non-contact in-process ultrasonic screening of thin fusion welded joints. *J. Manuf. Process.* **2021**, *64*, 445–454. [CrossRef]
45. Vasilev, M.; MacLeod, C.; Javadi, Y.; Pierce, G.; Gachagan, A. Feed forward control of welding process parameters through on-line ultrasonic thickness measurement. *J. Manuf. Process.* **2021**, *64*, 576–584. [CrossRef]

Article

Collaborative Robotic Wire + Arc Additive Manufacture and Sensor-Enabled In-Process Ultrasonic Non-Destructive Evaluation

Rastislav Zimermann [1,*], Ehsan Mohseni [1], Momchil Vasilev [1], Charalampos Loukas [1], Randika K. W. Vithanage [1], Charles N. Macleod [1], David Lines [1], Yashar Javadi [1], Misael Pimentel Espirindio E Silva [2], Stephen Fitzpatrick [2], Steven Halavage [2], Scott Mckegney [2], Stephen Gareth Pierce [3], Stewart Williams [3] and Jialuo Ding [3]

[1] Centre for Ultrasonic Engineering, University of Strathclyde, Glasgow G1 1XW, UK; ehsan.mohseni@strath.ac.uk (E.M.); momchil.vasilev@strath.ac.uk (M.V.); charalampos.loukas@strath.ac.uk (C.L.); randika.vithanage@strath.ac.uk (R.K.W.V.); charles.macleod@strath.ac.uk (C.N.M.); david.lines@strath.ac.uk (D.L.); yashar.javadi@strath.ac.uk (Y.J.)

[2] Advanced Forming Research Centre, University of Strathclyde, Renfrew PA4 9LJ, UK; misael.pimentel@strath.ac.uk (M.P.E.E.S.); s.fitzpatrick@strath.ac.uk (S.F.); steven.halavage@strath.ac.uk (S.H.); scott.mckegney@strath.ac.uk (S.M.)

[3] Welding Engineering and Laser Processing Centre, University of Cranfield, Cranfield MK43 0AL, UK; s.g.pierce@strath.ac.uk (S.G.P.); s.williams@cranfield.ac.uk (S.W.); jialuo.ding@cranfield.ac.uk (J.D.)

* Correspondence: rastislav.zimermann@strath.ac.uk

Citation: Zimermann, R.; Mohseni, E.; Vasilev, M.; Loukas, C.; Vithanage, R.K.W.; Macleod, C.N.; Lines, D.; Javadi, Y.; Espirindio E Silva, M.P.; Fitzpatrick, S.; et al. Collaborative Robotic Wire + Arc Additive Manufacture and Sensor-Enabled In-Process Ultrasonic Non-Destructive Evaluation. *Sensors* 2022, 22, 4203. https://doi.org/10.3390/s22114203

Academic Editor: Nachappa Gopalsami

Received: 4 April 2022
Accepted: 28 May 2022
Published: 31 May 2022

Publisher's Note: MDPI stays neutral with regard to jurisdictional claims in published maps and institutional affiliations.

Copyright: © 2022 by the authors. Licensee MDPI, Basel, Switzerland. This article is an open access article distributed under the terms and conditions of the Creative Commons Attribution (CC BY) license (https://creativecommons.org/licenses/by/4.0/).

Abstract: The demand for cost-efficient manufacturing of complex metal components has driven research for metal Additive Manufacturing (AM) such as Wire + Arc Additive Manufacturing (WAAM). WAAM enables automated, time- and material-efficient manufacturing of metal parts. To strengthen these benefits, the demand for robotically deployed in-process Non-Destructive Evaluation (NDE) has risen, aiming to replace current manually deployed inspection techniques after completion of the part. This work presents a synchronized multi-robot WAAM and NDE cell aiming to achieve (1) defect detection in-process, (2) enable possible in-process repair and (3) prevent costly scrappage or rework of completed defective builds. The deployment of the NDE during a deposition process is achieved through real-time position control of robots based on sensor input. A novel high-temperature capable, dry-coupled phased array ultrasound transducer (PAUT) roller-probe device is used for the NDE inspection. The dry-coupled sensor is tailored for coupling with an as-built high-temperature WAAM surface at an applied force and speed. The demonstration of the novel ultrasound in-process defect detection approach, presented in this paper, was performed on a titanium WAAM straight sample containing an intentionally embedded tungsten tube reflectors with an internal diameter of 1.0 mm. The ultrasound data were acquired after a pre-specified layer, in-process, employing the Full Matrix Capture (FMC) technique for subsequent post-processing using the adaptive Total Focusing Method (TFM) imaging algorithm assisted by a surface reconstruction algorithm based on the Synthetic Aperture Focusing Technique (SAFT). The presented results show a sufficient signal-to-noise ratio. Therefore, a potential for early defect detection is achieved, directly strengthening the benefits of the AM process by enabling a possible in-process repair.

Keywords: non-destructive evaluation; in-process robotic NDE; Wire + Arc Additive Manufacture (WAAM); ultrasound testing; total focusing method

1. Introduction

In 2019, the global metal Additive Manufacturing (AM) market size was valued at 2.02 billion € and was predicted to grow by up to 27.9% annually until 2024 [1]. AM technology plays a critical role in the latest industrial revolution, Industry 4.0, where there is a demand for smart factories capable of fabricating high-quality customized products [2].

One such AM technology, called Wire + Arc Additive Manufacturing (WAAM), is a rapidly developing metal AM technology, based on a directed energy deposition process [3], which promises an automated fabrication of structurally complicated three-dimensional (3D) near-net shaped components [4]. The process is aiming to achieve superior cost-efficiency by reducing energy usage, material waste and time as compared to traditional subtractive manufacturing methods [5]. The technology has attracted the attention of sectors such as aerospace and naval engineering, due to their interest in weight reduction and increased geometrical complexity of solid metal components [6]. Moreover, WAAM offers the potential to build products using otherwise expensive materials such as titanium alloys, steel or nickel-based super-alloys [4].

Conventionally, the quality assurance of WAAM is performed by Non-Destructive Evaluation (NDE) after the full built completion via manually deployed methods such as ultrasound testing [7], eddy-current [8] or X-ray based imaging [9]. These techniques, however, require complex and time taking manipulation between workstations and often pre-processing or machining of the components [9], affecting the production throughput and therefore overall cost if a defect is discovered. Hence, in order to maintain the benefits of the already highly automated WAAM process, the demand for automatically deployed flexible NDE integrated in-process is high [10].

The detection of defects, in-process, facilitates the potential for real-time repair or early scrapping of parts, preventing the manufacturer from time-taking deposition of costly material over defective layers. Moreover, the deployment of automated NDE offers greater benefits such as positional accuracy, repeatability and high rates of inspection as compared to human NDE operators [11].

Recently published research has presented an important advancement in the field of automated in-process NDE of arc-based welding processes [12–14], which are manufacturing methods with similar applicable attributes and challenges to WAAM. The development of a multi-robot welding cell demonstrated the possibility of robotic welding and automated ultrasound NDE [15]. Full automation was achieved by a novel sensor-enabled robotic system based around a real-time embedded controller which enabled: (a) real-time communication, (b) data acquisition and (c) control of the process. Moreover, a UDP (User Datagram Protocol) communication protocol established through the Robot System Interface (RSI) [16], developed by industrial robot manufacturer KUKA, was used for the sensor-based robotic motion correction that could influence the pre-programmed robot's path through the sensor's feedback on the fly. The motion corrections were executed based on a novel developed motion software operating in real-time intervals (4 milliseconds intervals for KUKA Robot Controller (KRC) 4). Therefore, it was reported possible to utilize the cell for automated ultrasound inspection in three modes: (1) post-process continuous, (2) inter-pass in-process or (3) live-arc in-process. Further, the use of Force-Torque (FT) sensor-driven robotic motion for automated NDE of complex geometries was explored in [10]. The FT sensor facilitated path correction required for contact-based scanning of the aircraft wing cover through an as-built surface geometry that was inconsistent with geometry in the original CAD model. Therefore, the research faced similar automation challenges, associated with transducer deployment on the estimated pre-programmed path, applicable to possible automated NDE deployment on near net-shaped WAAM.

The ultrasound-based in-process NDE of welds at elevated temperatures was made possible by the development of novel, phased array ultrasound transducer (PAUT)-based, high-temperature and dry-coupled roller-probes [15,17]. Thanks to its design, based on a PAUT coupled through a water delay line and a flexible silicone rubber, the roller-probe was reported capable of withstanding temperatures of up to 350 °C, which made this device well suited for in-process NDE of arc-based manufacturing processes, where the resistance to elevated temperatures is highly desired [17]. This was a significant advancement in ultrasound NDE transducer development, given a typical commercial PAUT can only operate up to around 60 °C, while commercial delay lines can offset this limit to the temperatures only up to around 150 °C for a short period of time [18].

Further, the roller-probe technology has also been developed to couple with an as-built surface of WAAM components without the use of liquid couplants [19]. However, when considering the roller-probe inspection approach, new challenges emerged, as the as-built WAAM component features a non-flat and varying surface geometry (in both the scanning and the traverse direction) resulting in high contact forces being required to assure full compliance of the roller-probe tyre to the surface. Hence, the design must facilitate the transmission of the maximum possible ultrasonic energy without suffering signal losses. This resulted in the alteration of the internal liquid delay line with a heat-resistant solid core (delay line) made of Polyetherimide Polymer.

The key advantage of the novel WAAM roller-probe NDE approach is the removal of the post-deposition processing stages, which often included waiting for the sample to cool down, machining operations and manipulation between workstations. The inspection would then be performed through a flat surface either using direct gel-coupled contact with the sample or in water immersion tanks using gantry systems [7,20,21].

Owing to the novel WAAM dry-coupled roller-probe ultrasound NDE concept, the research has presented a possibility to detect Lack of Fusion (LoF) defects as small as 5 × 0.5 × 0.5 mm (width, length and height), through an as-built surface of the WAAM wall [22]. The ultrasound data acquisition called Full Matrix Capture (FMC) enabled the collection of raw time-domain data without consideration of any refractive boundaries or couplant conditions as the imaging was executed at the post-processing stage [23]. The developed ultrasound post-processing algorithms, based on the Synthetic Aperture Focusing Technique (SAFT) and Total Focusing Method (TFM) made it possible to overcome complications associated with multiple refractions present at a non-flat surface of WAAM and internal components of the roller-probe [24]. These algorithms, also called the SAFT-TFM package, were based on Delay and Sum (DaS) computational logic, where at first the Time of Flight (ToF) elapsed, between a PAUT element and a targeted image pixel, was calculated. Subsequently, the signal response from the corresponding time sample of an elementary A-scan was summed to the pixel. ToF calculation was repeated for every transmit–pixel–receive combination, thus, a fully focused image of the WAAM interior was formed. This novel ultrasound NDE approach was, however, presented on static inspection of WAAM components and was not yet deployed in-process on high-temperature builds.

Therefore, in this paper, the authors present an experimental multi-robot cell designed for WAAM deposition and automated in-process dry-coupled ultrasound NDE using a custom WAAM roller-probe. Within the cell, the plasma-arc WAAM process was controlled by the deposition software while a full external control of the NDE process was achieved by the sensor-enabled adaptive motion control package adapted to in-process WAAM NDE. The automated high-temperature WAAM roller-probe was deployed within a dwell time, set for inter-layer cooling, while sufficient coupling with the as-built surface of WAAM during the inspection was assured by the FT sensor. In this work, a titanium WAAM straight component (wall) with embedded tungsten reflectors was deposited to evaluate the performance of the in-process NDE approach. The use of tungsten tubes as cylindrical artificial reflectors, with known diameter, for ultrasound inspection technique calibration and evaluation has found its application in the fields of in-process welding inspection [12,13] as well as ultrasound inspection of WAAM [25]. An advantage of the tungsten can be realized by the possibility to manufacture inclusions, closely simulating defects such as Lack of Fusion (LoF) or keyholes, at the desired location [26]. During the in-process NDE, the position encoded FMC data were acquired using a high-speed ultrasound phased array controller. The SAFT-TFM package, then, enabled the highly accurate detection of artificial reflectors presented on an amplitude C-scan image. C-scan imaging provided a top-view over an interior of the WAAM component and was found effective for data review from a large inspection volume [27]. Thus, for the first time, a volumetric in-process ultrasound NDE of as-built WAAM was achieved, directly supporting research on producing right-first-time WAAM parts.

2. The Architecture of the WAAM + NDE Cell

2.1. Hardware

The automated robotic WAAM and NDE system depicted in Figure 1 was designed based on 2 × 6 Degrees of Freedom (DoF) industrial robotic manipulators (KUKA KR90 R3100) employed as a WAAM deposition robot and as an inspection robot. Additionally, as a part of the deposition robot, a horizontal rotary positioner (KUKA DKP-400V3) was also located within this cell and utilized as a rotational tooling mainframe and substrate clamping device. The deposition hardware, physically mounted on a deposition robot's end-effector, featured a water-cooled plasma-arc welding torch (controlled by: EWM-TETRIX 552 AC/DC SYNERGIC PLASMA AW welder) integrated into a deposition device with a local shielding [28], as seen in Figure 1a. The local shielding device was an aluminum enclosure with multiple gas outlet channels fitted, that provided an additional supply of the argon shielding gas on a high-temperature WAAM to prevent atmospheric contamination that could result in oxidation of the fresh deposit. Further, a wire-feed outlet with adjustable height was fitted on the deposition device, positioned to supply feedstock into the melt pool. The wire supply was controlled by an automatic wire feeder (EWM T drive 4 Rob 3 Li, EWM) that was attached directly to the deposition robot's arm as well. Lastly, the deposition head was equipped with a high dynamic range welding camera (Xiris XVC-1000) used to remotely assess the deposition quality.

Figure 1. Implemented (**a**) WAAM deposition cell with plasma arc process, and (**b**) Roller-Probe based NDE.

An inspection robot, seen in Figure 1b was equipped with an FT sensor (FTN-GAMMA-IP65 SI-130-10, Schunk, Germany) mounted on the end effector. A WAAM roller-probe was, then, attached to an FT sensor serving as an end effector to the robot flange. The roller-probe, depicted in detail on Figure 2, was driven by a high-speed phased array ultrasound controller LTPA (PEAK NDT, United Kingdom) mounted directly on the robot arm. Further, the communication between all hardware was achieved by a network switch (Zyxel Gigabit ethernet switch) enabling control of the WAAM process and NDE via a single ethernet connection plugged into the PC.

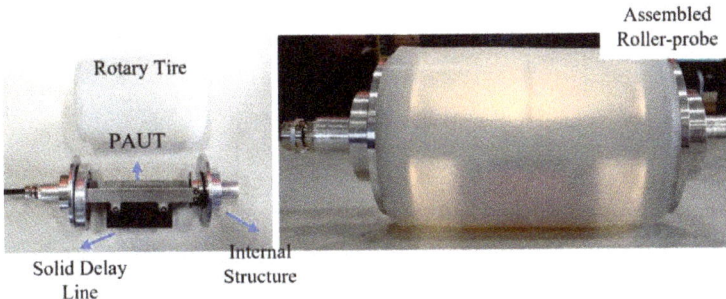

Figure 2. The internal structure of the roller-probe (**left**) and assembled device (**right**).

2.2. Software Setup

2.2.1. Deposition

In this work, the deposition robot was controlled by a pre-installed PC with a WAAM-Ctrl (WAAM3D, Milton Keynes, UK) [29] application, streaming the deposition commands (robot paths, deposition parameters) directly to the deposition robot via RSI over an ethernet connection. The tool-path plan was generated using WAAMPlanner Software (WAAM3D, UK) [30], where the desired component was imported as a Computer-Aided Design (CAD), sliced into layers according to the pre-defined layer height, segmented into a set of individual building blocks from which the series of tool-paths was generated. Depending on the variables, such as material, geometry or deposition process, the deposition parameters were given to a WAAMPlanner and the post-processed file was generated, translating the information to a ready-to-stream xml file.

2.2.2. NDE Software

The NDE inspection was guided by a software platform developed in the LabVIEW programming environment [31], which offers reliable communication between instruments and fast prototyping, through several available toolboxes and libraries. The Graphic User Interface (GUI) is presented in Figure 3, where the platform consisted of parallel state machines responsible for executing the program in sequence, controlling the inspection robot kinematics through the FT sensor feedback and ultrasound data acquisition in real-time.

The real-time corrections (every 4 milliseconds) of the robot's motion, based on linear interpolation, and control used for the in-process NDE work were based on a flexible robotic motion framework presented in [15] and developed for in-process inspection and automated NDE purposes. During the inspection, real-time adjustments of the inspection robot velocity, acceleration and contact force were available. Position-determined triggers were implemented to automatically switch between inspection and travel speed of the inspection robot, enabling/disabling the FT sensor-driven motion and data acquisition when needed. The Z-axis force control through the FT sensor was used to maintain sufficient contact with the WAAM component while the operator maintained the ability of real-time adjustments of the kinematics. In this work, the Z-axis motion corrections, associated with maintaining a steady force at the inspection speed, were calculated by the KRC based on the RSI configuration diagram. The X and Y translation, and A, B and C rotation-axes motion correction always remained in control of the initial inspection path-planning, while the appropriate motion corrections were calculated within the developed motion framework in 4 ms intervals and streamed through the RSI.

Further, taking the advantage of real-time communication with the inspection robot, the timestamped position of the inspection robot during an inspection was encoded to each FMC frame acquired. The FMC data were then processed within a MATLAB environment using a SAFT-TFM algorithm package, enabling positionally accurate analysis.

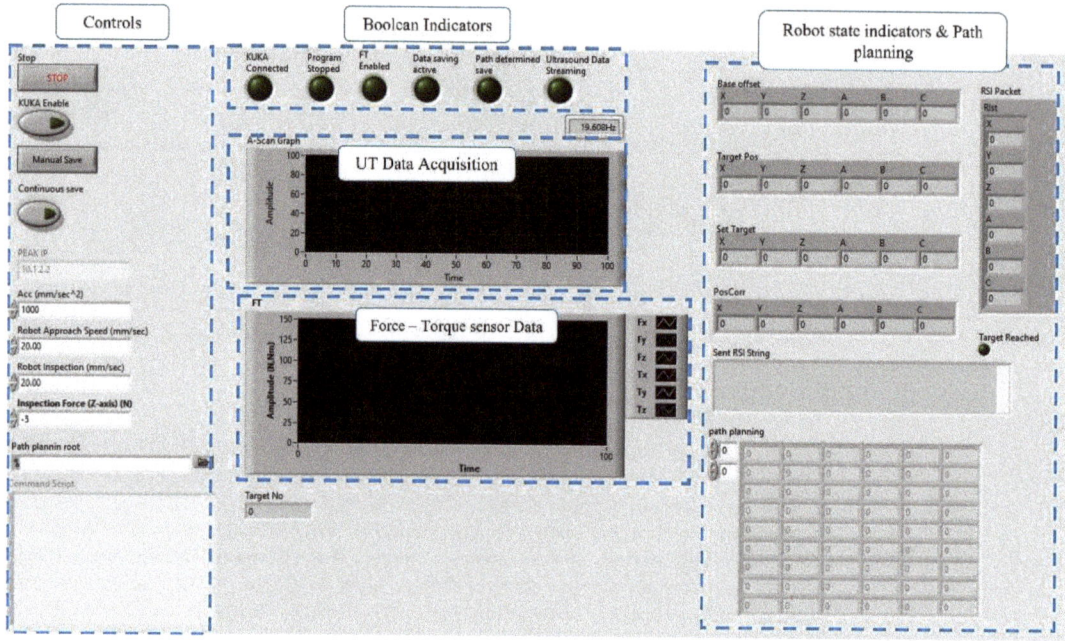

Figure 3. LabVIEW GUI for NDE process control and monitoring.

3. Experimental WAAM Manufacturing

3.1. WAAM Wall Path Planning and Deposition Parameters

To demonstrate the WAAM and NDE cell concept, and evaluate its performance, a titanium (Ti-6Al-4V) WAAM wall was chosen and designed for fabrication. The experimental wall was set to be 300.0 mm in length, 25.0 mm wide and a height given as 25.0 mm. However, knowing the nature of the WAAM process delivering near net-shaped components [4], extra material volume post-deposition was expected. Moreover, the height of the wall was not considered important since the goal was to evaluate the inspection of WAAM's interior with a specific volume. Therefore, the built process was stopped when the wall was found sufficiently high for in-process NDE demonstration to be performed.

The path planning designed in WAAMPlanner, seen in Figure 4, consisted of an oscillating deposition strategy [32], where a single bead, with a square zig-zag pattern, was deposited per layer. Relevant deposition parameters can be seen in Table 1 below.

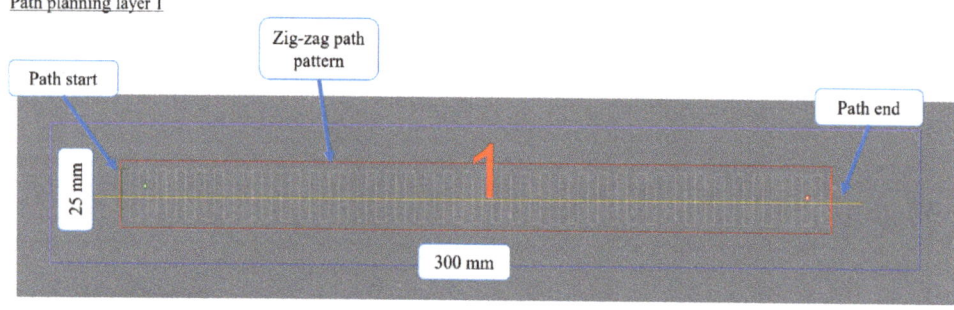

Figure 4. Deposition Path Planning for Layer 1 of an experimental WAAM wall.

Table 1. Deposition Parameters.

Deposition Parameters	
Current	150 Amps
Wire-feed speed	2.5 m/min
Robot Velocity	0.005 m/s

3.2. WAAM Wall Deposition

Figure 5a displays a deposition setup where an experimental wall was built on a Ti-6V-4AL substrate plate, 12.0 mm thick, clamped to the tooling which was mounted on a rotary table of the horizontal positioner. The plate was clamped using welding clamps to prevent bending caused by heat-induced residual stress [33], typical for arc-based manufacturing processes such as welding [34].

a) Substrate plate clamping with 1st deposited layer

b) Active Deposition

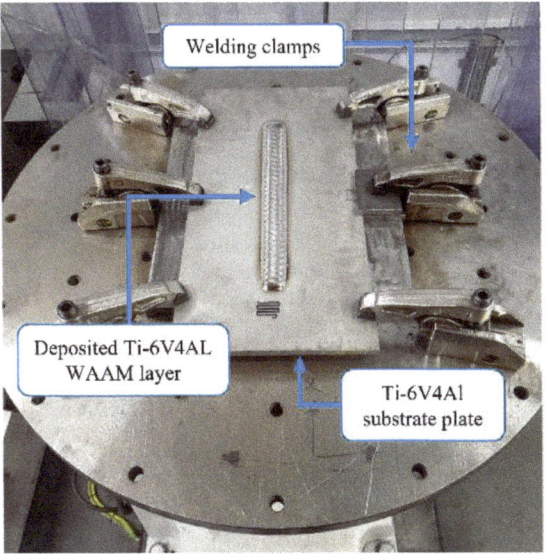
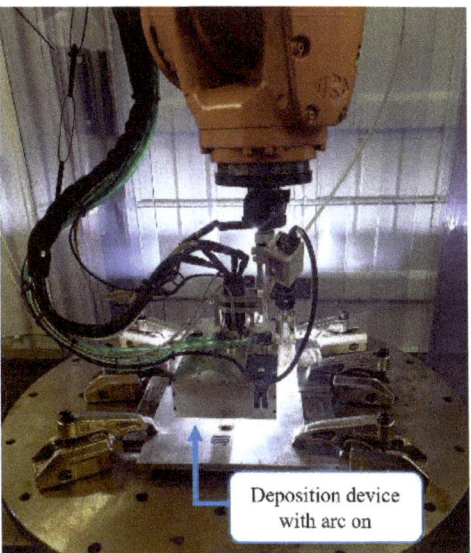

Figure 5. Deposition clamping setup and a substrate plate with a deposited 1st layer (**a**) and deposition process with an active torch (**b**).

This clamping set-up has created a challenging and restricting working envelope; hence, the first stage of manufacturing was calibration and verification of the path motion by a dry run. At this step of WAAM part fabrication, the robot traveled through the produced deposition paths without an active torch or wire feed. Therefore, the correct positioning of the robot could be assured, knowing that the deposition head would not collide with the clamping. This was extremely important, especially during the deposition of the first few layers, after which the deposition head was high enough not to collide with welding clamps.

Following, Figure 5b shows an active deposition of the 1st layer, while the completed pass on the substrate plate is visible in Figure 5a image. It is worth mentioning, that the height of the first layer was measured to be 3.5 mm.

3.3. Ultrasound Reflector Planting

To evaluate the NDE defect detection capability, artificial reflectors were embedded into the experimental wall. In this work, tungsten tubes with parameters specified in Table 2 were utilized for this purpose. Two tubes were embedded into layer 3 by producing slots using a portable grinding machine. The tubes were located approximately 55 mm from each other. Tube 1 was placed parallel to the wall, in the approximate centre of the bead. Tube 2, on the other hand, was embedded in the transverse direction to the wall as seen in Figure 6a.

Table 2. Tungsten Tube parameters.

Tungsten Tube	
Tube length	30 mm
Internal diameter	1 mm
Outer diameter	3 mm

a) Tungsten tube placement into layer 3

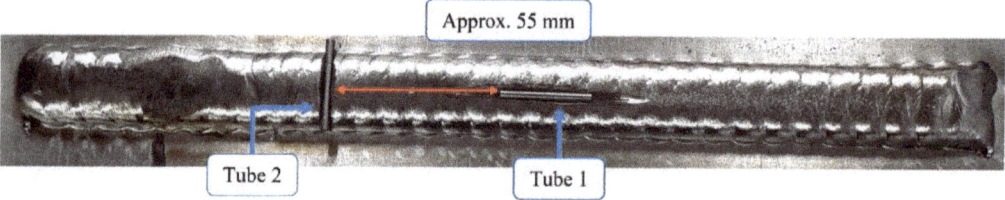

b) Fully covered tungsten tubes after pass 4

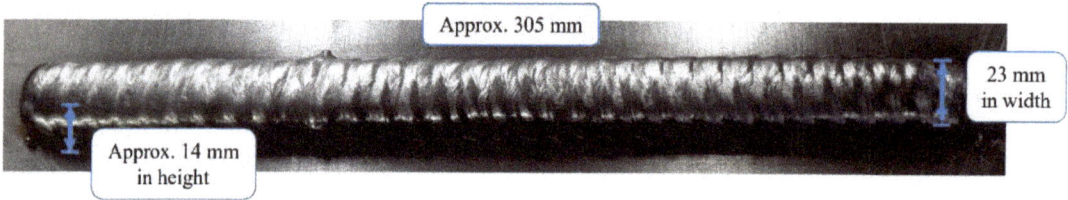

Figure 6. Tungsten tube embedding into layer 3 (**a**) and a subsequently deposited layer 4 covering tubes (**b**).

Further, Figure 6b depicts the wall after layer 4 where the tungsten rods were fully covered by the freshly deposited titanium. No significant inconsistencies (defects) in the surface quality that could cause a potential failure of the building process were observed once layer 4 was completed. However, a minor material built up was observed which was corrected after the subsequent layer deposition.

4. In-Process NDE of the Experimental WAAM Wall

4.1. Ultrasound Inspection Parameters

The ultrasound data were acquired using a roller-probe featuring a solid delay line housed in a silicone rubber tyre. The PAUT, with specifications found in Table 3, was positioned to sit on the top of the delay line.

Table 3. PAUT parameters.

Array Parameters	Value
Element Count	64
Element Pitch	0.5 mm
Element Elevation	10 mm (unfocused)
Element Spacing	0.1 mm
Centre Frequency	5 MHz

The FMC data were collected using an LTPA phased array controller with 200 V excitation voltage and a fixed hardware gain of 60 dB. The time-domain matrix of the signals was formed by 3000 data samples for each transmits–receive pair at a sampling frequency of 50 MHz. During the data post-processing stage, the following acoustic velocities for longitudinal ultrasound waves were used for refraction and time-of-flight computations: (1) Delay line = 2480 m/s, (2) Rubber = 1006 m/s and (3) Titanium = 6100 m/s. These values were obtained by ultrasound pulse-echo measurements of the individual roller–probe's components and titanium coupons cut from a previous trial and heated to 150 °C.

4.2. In-Process NDE

To demonstrate the ultrasound in-process NDE capability on the titanium wall with embedded tungsten tubes, producing an air gap inside the WAAM, two subsequent layers were deposited to build a six-layer-high component. The deposition of two additional layers enabled a natural surface profile common for plasma WAAM deposition [4], without any significant negative influence from previous grinding and tungsten tube embedding.

Figure 7 shows a completed deposition of the experimental wall after layer six with a measured height of approximately 21 mm with an average layer height of approximately 3.5 mm. A width of 28 mm and a length of 305 mm were also measured.

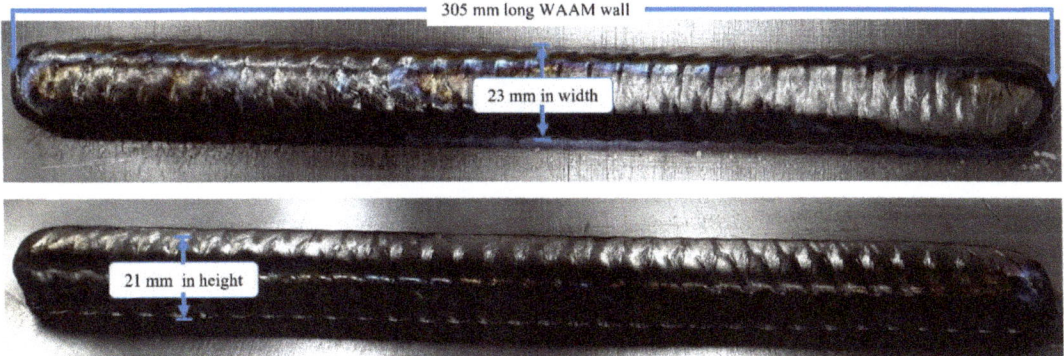

Figure 7. Completed experimental wall and its dimensions.

As suggested by the literature [32], there is an optimal inter-pass dwelling time to allow for cooling of Ti-6Al-4V WAAM built using the oscillation deposition strategy. Therefore, the in-process NDE can be integrated into the build process to leverage this inter-pass dwelling time to complete the inspection of the last pass without delaying the built process. Accordingly, a dwell time of 9 min was set to allow inter-pass cooling during the deposition of the experimental wall as suggested by [35]. This time was set to avoid the formation of coarse α_{GB} phase grain microstructure, and thus, achieve optimal mechanical properties of this hypothetical component. Moreover, the time was found sufficiently long for in-process NDE to be performed without causing costly delays in the production process.

Before starting the NDE, the surface temperature was taken using a handheld thermometer. The surface temperature of the WAAM was measured and ranged to be between 180–230 °C along the wall, which was much lower than the operational limit of the roller-probe (resistant up to 350 °C).

The NDE was initiated within the first 2 min of the deposition robot's retraction to its home position. Figure 8 shows a step-by-step inspection diagram, where at first, the inspection robot's end-effector approached the wall with a travel speed of 50 mm/s until the position 5 mm above the predicted as-built surface of WAAM was reached.

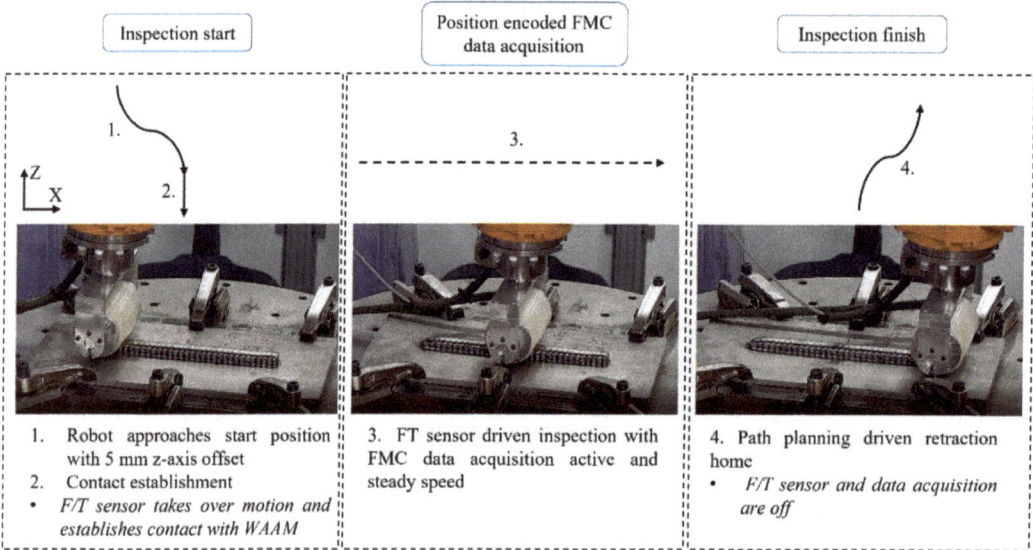

Figure 8. Inspection diagram showing the process and the sequence of robotic motions during an inspection.

The second stage in the diagram shows a contact establishment with the WAAM specimen. This was accomplished by an automatic trigger that recognized the robot's position (5 mm above the expected surface), which was followed by a change of robot speed to an inspection speed (in this work = 2 mm/s) and initiation of FT sensor-driven motion. A command to maintain a constant force of 130 N was sent to the inspection robot from LabVIEW via RSI; thus, the Z-axis position correction was no longer managed by a LabVIEW motion framework, but the kinematics corrections were calculated and applied by the KRC. The force applied to the component was set to a maximum force given by the FT sensor operational limit.

During the descending of the inspection robot on the surface of WAAM, the LabVIEW program was set to wait for 2 s, before sending coordinates of the next position. This "wait" command enabled the inspection robot to position itself on the surface with the required set stable force and without further freedoms in the X-Y plane that could result in inconsistent contact with a specimen.

Stage 3 of the inspection was initiated by sending coordinates of the next target position (in this scenario = the end of an inspection, +300 mm in the X-axis direction) and enabling encoded FMC data acquisition. The FMC data were acquired while the inspection robot traveled along the path with a steady force by correcting its Z-axis position to maintain a given force value with the experimental wall.

Once the end of the path was reached, the termination of the inspection was triggered by the change of the inspection robot's Z-axis targeted position. This was given as the Z-axis target position offset larger than 5 mm above the predicted WAAM surface. The trigger was used to disable the sensor-driven motion and the ultrasound data acquisition. The process was concluded by retracting the inspection robot to its home position according to the path planning.

The inspection volume from the experimental wall was set to 300 mm, therefore the time elapsed to inspect the component equaled 150 s with an additional approximate 60 s that included the approach to the specimen and the robot retraction back to a home position. It is worth mentioning, that the entire inspection took significantly less time than the period set for a dwelling (9 min), which complemented the objectives required for the in-process NDE of WAAM in this scenario. The total number of positions encoded FMC frames acquired was 200, giving a sample density of 1.5 per mm (sampling frequency = 0.75 Hz).

4.3. Ultrasound Data Post-Processing: TFM Imaging and C-Scan

After the completion of the in-process NDE, the ultrasound data were processed using a SAFT-TFM algorithm described in [22]. The TFM frames (B-Scan) were computed for a 25 mm × 19 mm region at 6 pixels/mm resolution, which is compatible with the 2 dB Amplitude Fidelity criterion of ASME V [36]. This window represented an internal volume of the desired component between the baseplate and a region 2 mm beneath the surface or just above the interface of layers 5 and 6, where potential defects would be expected. Moreover, this work was focused on the detection of tungsten tubes, therefore there was no interest in detecting and analyzing possible generated true defects from the WAAM process, since the calibration for these defects has not yet been developed.

To achieve a full C-scan, the computation was initiated by the ultrasound surface reconstruction using a SAFT surface imaging and surface finding algorithm. Afterward, the curves representing the WAAM surface contours were augmented into the 3-layer adaptive TFM algorithm to produce the TFM frames before their normalization. Normalizing all the frames used to construct the C-scan aided to visualize the entire image on the same dB scale. Using the raw unnormalized frames, the C-scan was formed by populating a new 2-dimensional array's columns with maximum detected amplitudes from all TFM frame's columns from n number (n = 200) of TFM frames.

The size of the C-scan presented in this paper was set to 150 × 200 pixels (Number of pixels in the horizontal axis of the TFM frame × the number of frames). The resulting C-scan image was normalized and plotted on a dB scale from the peak amplitude to an averaged noise level (0 to −12 dB), giving the best visual contrast between a signal from tungsten tubes and interference from the base noise levels.

5. In-Process Inspection Results and Discussion

In this section, the results of an in-process NDE are presented and discussed. The outcome of the in-process inspection is depicted in Figure 9a, where the signal from Tube 1 and Tube 2, with an internal diameter of 1.0 mm was successfully detected. At a first glance, stronger signal levels are observed from a longitudinally placed 30 mm long Tube 1. Noteworthy that a matching signal extension of approximately 30 mm along the inspection travel direction is also well noticeable. Tube 2, embedded in the traverse direction, shows visually weaker signal strength where the energy from the tube is represented by a concentrated signal in the centre of the corresponding frames approximately 100 mm from the inspection start point.

Following a visual analysis of the results, a maximum amplitude along an X-axis was presented in Figure 9b. Based on this plot, a Signal-to-Noise ratio (SNR) of up to 12 dB was achieved from the scanning of Tube 1 while an SNR of 10 dB was seen for Tube 2. Considering the dry-coupling condition, these SNR values were found sufficiently high for the indications to stand out from the background noise and be readily detected by the operator.

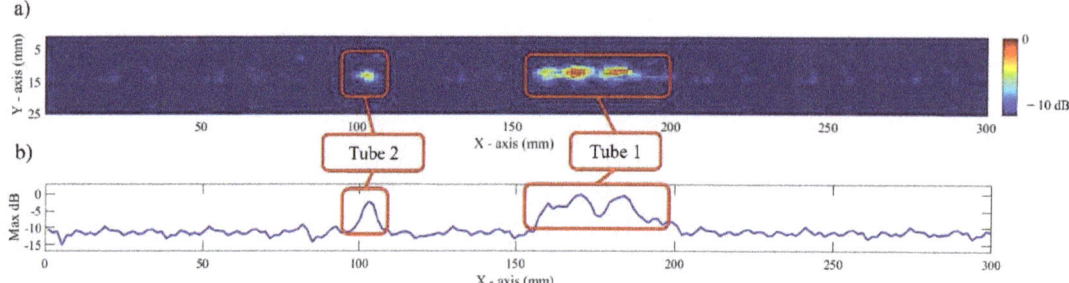

Figure 9. Results showing: (**a**) C-scan obtained from computed TFM frames and (**b**) extracted maximum amplitude along X-axis.

Further analysis shows signal strength variations from Tube 1 signal along the scan path where an SNR drop of only 4 dB was observed. This local signal strength loss can be associated with the possible changes to the contact quality between the rubber tyre and the non-flat wave-like surface profile of WAAM. This means the signal strength propagating into the specimen was fluctuating with the varying profile of WAAM. Further losses of SNR, especially for Tube 2, could be associated with a lack of compensation for the thermal gradient that affects an ultrasound wave velocity during propagation, as also pointed out in [12,37]. This means the image signal amplitude is negatively affected due to the loss of focusing precision during TFM image forming.

6. Conclusions and Future Work

In this paper, a design and demonstration of a novel multi-robot cell for WAAM and ultrasound in-process NDE was presented. The architecture, based on two robotic industrial manipulators featuring a deployed plasma arc WAAM process and high-temperature PAUT roller-probe was introduced along with a software control package, merging manufacture and NDE into a single continuous process.

The in-process NDE capability was demonstrated on a dry-coupled ultrasound in-process inspection of the Ti-6Al-4V WAAM wall with embedded tungsten tube reflectors, with an internal diameter of 1.0 mm. Using the FMC data acquisition, a C-scan image of the experimental wall was computed by deploying a SAFT-TFM package. The results of the in-process inspection showed successfully detected embedded tubes, with distinguishable SNR of up to 12 dB.

Therefore, this work demonstrated the ability to detect defects just after the point of generation, which can pave the way for possible in-process repair processes to be deployed in the future. It can also be concluded that the presented research enables the further amplification of WAAM benefits by the deployment of flexible and automated NDE.

The future development is aimed at improving the speed of image forming, by the employment of graphics processing units. Moreover, the performance evaluation is targeted at transduction and automated deployment in various scenarios such as (1) inspection of geometrically complex WAAM components, (2) an investigation of probe deployment while the torch is active elsewhere and (3) at varying robotic NDE speeds. For the in-process defect detection and characterization area, the key aims are based on the development of thermal gradient compensation capabilities, which can further enhance defect detection and accuracy of the inspection approach.

Lastly, the research aims to develop a defect calibration procedure for various materials and a wide range of natural defects to enable automated defect detection and characterization that can further enhance the automation of the WAAM.

Author Contributions: Conceptualization, R.Z., E.M. and C.N.M.; methodology, R.Z., E.M. and R.K.W.V.; software, R.Z., M.V. and C.L.; validation, R.Z., E.M., Y.J., S.M. and D.L.; formal analysis, R.Z. and E.M.; investigation, R.Z., S.H. and S.M.; resources, S.F., S.W., C.N.M., S.G.P., M.P.E.E.S. and J.D.; data curation, R.Z., E.M. and D.L.; writing—original draft preparation, R.Z. and E.M.; writing—review and editing, E.M., R.K.W.V., S.W., D.L., C.N.M., M.V., C.L. and S.G.P.; visualization, E.M., D.L. and R.Z; supervision, C.N.M., S.W., S.G.P., J.D. and S.F.; project administration, S.W., C.N.M., S.G.P. and S.F.; funding acquisition, C.N.M., S.G.P. and S.W. All authors have read and agreed to the published version of the manuscript.

Funding: This research was funded by EPSRC: (I) NEWAM (EP/R027218/1), (II) EPSRC Doctoral Training Partnership (DTP) (EP/R513349/1) and RoboWAAM (III) EP/P030165/1.

Institutional Review Board Statement: Not applicable.

Informed Consent Statement: No applicable.

Data Availability Statement: The data can be shared upon reasonable request.

Acknowledgments: The project was supported by EPSRC: (I) NEWAM (EP/R027218/1) and (II) EPSRC Doctoral Training Partnership (DTP) (EP/R513349/1) and RoboWAAM (III) EP/P030165/1. Further, the authors would like to thank the technical staff at the Lightweight Manufacturing Centre (LMC) (Renfrew, UK) for the support of this work, and the team at KUKA Robotics (Jeff Nowill, Alan Oakley, Steve Hudson in particular) for enabling and supporting the robotics work.

Conflicts of Interest: The Authors declare no conflict of interests.

References

1. Metal Additive Manufacture Market. Available online: https://additive-manufacturing-report.com/additive-manufacturing-market/ (accessed on 17 January 2022).
2. Zawadzki, P.; Żywicki, K. Smart product design and production control for effective mass customization in the Industry 4.0 concept. *Manag. Prod. Eng. Rev.* **2016**, *7*, 105–112. [CrossRef]
3. DebRoy, T.; Mukherjee, T.; Milewski, J.; Elmer, J.; Ribic, B.; Blecher, J.; Zhang, W. Scientific, technological and economic issues in metal printing and their solutions. *Nat. Mater.* **2019**, *18*, 1026–1032. [CrossRef] [PubMed]
4. Williams, S.W.; Martina, F.; Addison, A.C.; Ding, J.; Pardal, G.; Colegrove, P. Wire+ arc additive manufacturing. *Mater. Sci. Technol.* **2016**, *32*, 641–647. [CrossRef]
5. Wu, B.; Pan, Z.; Ding, D.; Cuiuri, D.; Li, H.; Xu, J.; Norrish, J. A review of the wire arc additive manufacturing of metals: Properties, defects and quality improvement. *J. Manuf. Processes* **2018**, *35*, 127–139. [CrossRef]
6. Thapliyal, S. Challenges associated with the wire arc additive manufacturing (WAAM) of aluminum alloys. *Mater. Res. Express* **2019**, *6*, 112006. [CrossRef]
7. Mohseni, E.; Javadi, Y.; Lines, D.; Vithanage, W.; Kosala, R.; Foster, E.; Qiu, Z.; Zimermann, R.; MacLeod, C.N.; Pierce, S. Ultrasonic phased array inspection of wire plus arc additive manufactured (WAAM) titanium samples. In Proceedings of the 58th Annual British Conference on Non-Destructive Testing, Telford, UK, 3–5 November 2019.
8. García-Martín, J.; Gómez-Gil, J.; Vázquez-Sánchez, E. Non-destructive techniques based on eddy current testing. *Sensors* **2011**, *11*, 2525–2565. [CrossRef]
9. Lopez, A.; Bacelar, R.; Pires, I.; Santos, T.G.; Sousa, J.P.; Quintino, L. Non-destructive testing application of radiography and ultrasound for wire and arc additive manufacturing. *Addit. Manuf.* **2018**, *21*, 298–306. [CrossRef]
10. Mineo, C.; MacLeod, C.; Morozov, M.; Pierce, S.G.; Lardner, T.; Summan, R.; Powell, J.; McCubbin, P.; McCubbin, C.; Munro, G. Fast ultrasonic phased array inspection of complex geometries delivered through robotic manipulators and high speed data acquisition instrumentation. In Proceedings of the 2016 IEEE International Ultrasonics Symposium (IUS), Tours, France, 18–21 September 2016; pp. 1–4.
11. Kaczmarek, M.; Piwakowski, B.; Drelich, R. Noncontact ultrasonic nondestructive techniques: State of the art and their use in civil engineering. *J. Infrastruct. Syst.* **2017**, *23*, B4016003. [CrossRef]
12. Javadi, Y.; Sweeney, N.E.; Mohseni, E.; MacLeod, C.N.; Lines, D.; Vasilev, M.; Qiu, Z.; Vithanage, R.K.; Mineo, C.; Stratoudaki, T. In-process calibration of a non-destructive testing system used for in-process inspection of multi-pass welding. *Mater. Des.* **2020**, *195*, 108981. [CrossRef]
13. Javadi, Y.; Mohseni, E.; MacLeod, C.N.; Lines, D.; Vasilev, M.; Mineo, C.; Foster, E.; Pierce, S.G.; Gachagan, A. Continuous monitoring of an intentionally-manufactured crack using an automated welding and in-process inspection system. *Mater. Des.* **2020**, *191*, 108655. [CrossRef]
14. Javadi, Y.; Macleod, C.; Lines, D.; Vasilev, M.; Mohseni, E.; Foster, E.; Qiu, Z.; Vithanage, R.; Zimermann, R.; Loukas, C. In-process inspection of multi-pass robotic welding. In Proceedings of the 46th Annual Review of Progress in Quantitative Nondestructive Evaluation, Portland, OR, USA, 14–18 July 2019.

15. Vasilev, M.; MacLeod, C.N.; Loukas, C.; Javadi, Y.; Vithanage, R.K.; Lines, D.; Mohseni, E.; Pierce, S.G.; Gachagan, A. Sensor-enabled multi-robot system for automated welding and in-process ultrasonic nde. *Sensors* **2021**, *21*, 5077. [CrossRef] [PubMed]
16. KUKA.RobotSensorInterface 4.0. 2018. Available online: https://www.kuka.com/en-gb/products/robotics-systems/software/system-software/kuka_systemsoftware (accessed on 19 January 2022).
17. Vithanage, R.K.; Mohseni, E.; Qiu, Z.; MacLeod, C.; Javadi, Y.; Sweeney, N.; Pierce, G.; Gachagan, A. A phased array ultrasound roller probe for automated in-process/interpass inspection of multipass welds. *IEEE Trans. Ind. Electron.* **2020**, *68*, 12781–12790. [CrossRef]
18. Olympus-ims.com. Ultrasonic Phased Array Wedge for Inspecting High-Temperature Parts up to 150 °C. Available online: https://www.olympus-ims.com/en/applications/ultrasonic-phased-array-wedge-for-inspecting-high-temperature-parts-up-to-150c/#:~{}:text=As%20phased%20array%20probes%20heat,temperature%20of%2060%20%C2%B0C (accessed on 16 November 2020).
19. Mohseni, E.; Vithanage, R.K.W.; Qiu, Z.; MacLeod, C.N.; Javadi, Y.; Lines, D.; Zimermann, R.; Pierce, G.; Gachagan, A.; Ding, J. A high temperature phased array ultrasonic roller probe designed for dry-coupled in-process inspection of wire+ arc additive manufacturing. In Proceedings of the 47th Annual Review of Progress in Quantitative Nondestructive Evaluation, Virtual, 25–26 August 2020.
20. Lopez, A.B.; Santos, J.; Sousa, J.P.; Santos, T.G.; Quintino, L. Phased Array Ultrasonic Inspection of Metal Additive Manufacturing Parts. *J. Nondestruct. Eval.* **2019**, *38*, 62. [CrossRef]
21. Chauveau, D. Review of NDT and process monitoring techniques usable to produce high-quality parts by welding or additive manufacturing. *Weld. World* **2018**, *62*, 1097–1118. [CrossRef]
22. Zimermann, R.; Mohseni, E.; Lines, D.; Vithanage, R.K.W.; MacLeod, C.N.; Pierce, S.G.; Gachagan, A.; Javadi, Y.; Williams, S.; Ding, J. Multi-Layer Ultrasonic Imaging of As-Built Wire + Arc Additive Manufactured Components. *Addit. Manuf.* **2021**, *48*, 102398. [CrossRef]
23. Holmes, C.; Drinkwater, B.; Wilcox, P. The post-processing of ultrasonic array data using the total focusing method. *Insight-Non Destr. Test. Cond. Monit.* **2004**, *46*, 677–680. [CrossRef]
24. Zimermann, R.; Mohseni, E.; Wathavana Vithanage, R.K.; Lines, D.; MacLeod, C.N.; Pierce, G.; Gachagan, A.; Williams, S.; Ding, J.; Marinelli, G. Implementation of an ultrasonic total focusing method for inspection of unmachined wire + arc additive manufacturing components through multiple interfaces. In Proceedings of the 47th Annual Review of Progress in Quantitative Nondestructive Evaluation, Virtual, 25–26 August 2020.
25. Javadi, Y.; MacLeod, C.N.; Pierce, S.G.; Gachagan, A.; Lines, D.; Mineo, C.; Ding, J.; Williams, S.; Vasilev, M.; Mohseni, E. Ultrasonic phased array inspection of a Wire+ Arc Additive Manufactured (WAAM) sample with intentionally embedded defects. *Addit. Manuf.* **2019**, *29*, 100806. [CrossRef]
26. Mohseni, E.; Javadi, Y.; Sweeney, N.E.; Lines, D.; MacLeod, C.N.; Vithanage, R.K.; Qiu, Z.; Vasilev, M.; Mineo, C.; Lukacs, P. Model-assisted ultrasonic calibration using intentionally embedded defects for in-process weld inspection. *Mater. Des.* **2021**, *198*, 109330. [CrossRef]
27. Mineo, C.; MacLeod, C.; Morozov, M.; Pierce, S.G.; Summan, R.; Rodden, T.; Kahani, D.; Powell, J.; McCubbin, P.; McCubbin, C. Flexible integration of robotics, ultrasonics and metrology for the inspection of aerospace components. *Proc. AIP Conf. Proc.* **2017**, *1806*, 020–026.
28. Xu, X.; Ding, J.; Ganguly, S.; Diao, C.; Williams, S. Preliminary investigation of building strategies of maraging steel bulk material using wire+ arc additive manufacture. *J. Mater. Eng. Perform.* **2019**, *28*, 594–600. [CrossRef]
29. WAAMctrl. Available online: https://waam3d.com/software/control-suite (accessed on 18 January 2022).
30. WAAMPlanner. Available online: https://waam3d.com/software/planner (accessed on 17 January 2022).
31. "What is LabVIEW?". Available online: https://www.ni.com/en-gb/shop/labview.html (accessed on 17 January 2022).
32. Aldalur, E.; Veiga, F.; Suárez, A.; Bilbao, J.; Lamikiz, A. High deposition wire arc additive manufacturing of mild steel: Strategies and heat input effect on microstructure and mechanical properties. *J. Manuf. Processes* **2020**, *58*, 615–626. [CrossRef]
33. Colegrove, P.A.; Coules, H.E.; Fairman, J.; Martina, F.; Kashoob, T.; Mamash, H.; Cozzolino, L.D. Microstructure and residual stress improvement in wire and arc additively manufactured parts through high-pressure rolling. *J. Mater. Process. Tech.* **2013**, *213*, 1782–1791. [CrossRef]
34. Javadi, Y.; Akhlaghi, M.; Najafabadi, M.A. Using finite element and ultrasonic method to evaluate welding longitudinal residual stress through the thickness in austenitic stainless steel plates. *Mater. Des.* **2013**, *45*, 628–642. [CrossRef]
35. Vázquez, L.; Rodríguez, N.; Rodríguez, I.; Alberdi, E.; Álvarez, P. Influence of interpass cooling conditions on microstructure and tensile properties of Ti-6Al-4V parts manufactured by WAAM. *Weld. World* **2020**, *64*, 1377–1388. [CrossRef]
36. Lines, D.; Mohseni, E.; Zimermann, R.; Mineo, C.; MacLeod, C.N.; Pierce, G.; Gachagan, A. Modelling of echo amplitude fidelity for transducer bandwidth and TFM pixel resolution. In Proceedings of the 47th Annual Review of Progress in Quantitative Nondestructive Evaluation, Virtual, 25–26 August 2020.
37. Slongo, J.S.; Gund, J.; Passarin, T.A.R.; Pipa, D.R.; Ramos, J.E.; Arruda, L.V.; Junior, F.N. Effects of Thermal Gradients in High-Temperature Ultrasonic Non-Destructive Tests. *Sensors* **2022**, *22*, 2799. [CrossRef] [PubMed]

Article

A Novel Complete-Surface-Finding Algorithm for Online Surface Scanning with Limited View Sensors

Alastair Poole [1,2,*], Mark Sutcliffe [2], Gareth Pierce [1] and Anthony Gachagan [1]

1 Centre of Ultrasonic Engineering (CUE), University of Strathclyde, Glasgow G1 1XW, UK; s.g.pierce@strath.ac.uk (G.P.); a.gachagan@strath.ac.uk (A.G.)
2 The Welding Institute (TWI) Wales, Port Talbot SA13 1SB, UK; mark.sutcliffe@twi.co.uk
* Correspondence: alastair.poole@strath.ac.uk

Abstract: Robotised Non-Destructive Testing (NDT) has revolutionised the field, increasing the speed of repetitive scanning procedures and ability to reach hazardous environments. Application of robot-assisted NDT within specific industries such as remanufacturing and Aerospace, in which parts are regularly moulded and susceptible to non-critical deformation has however presented drawbacks. In these cases, digital models for robotic path planning are not always available or accurate. Cutting edge methods to counter the limited flexibility of robots require an initial pre-scan using camera-based systems in order to build a CAD model for path planning. This paper has sought to create a novel algorithm that enables robot-assisted ultrasonic testing of unknown surfaces within a single pass. Key to the impact of this article is the enabled autonomous profiling with sensors whose aperture is several orders of magnitude smaller than the target surface, for surfaces of any scale. Potential applications of the algorithm presented include autonomous drone and crawler inspections of large, complex, unknown environments in addition to situations where traditional metrological profiling equipment is not practical, such as in confined spaces. In simulation, the proposed algorithm has completely mapped significantly curved and complex shapes by utilising only local information, outputting a traditional raster pattern when curvature is present only in a single direction. In practical demonstrations, both curved and non-simple surfaces were fully mapped with no required operator intervention. The core limitations of the algorithm in practical cases is the effective range of the applied sensor, and as a stand-alone method it lacks the required knowledge of the environment to prevent collisions. However, since the approach has met success in fully scanning non-obstructive but still significantly complex surfaces, the objectives of this paper have been met. Future work will focus on low-accuracy environmental sensing capabilities to tackle the challenges faced. The method has been designed to allow single-pass scans for Conformable Wedge Probe UT scanning, but may be applied to any surface scans in the case the sensor aperture is significantly smaller than the part.

Keywords: NDT; free-form surface profiling; autonomous robotic systems

Citation: Poole, A.; Sutcliffe, M.; Pierce, G.; Gachagan, A. A Novel Complete-Surface-Finding Algorithm for Online Surface Scanning with Limited View Sensors. *Sensors* **2021**, *21*, 7692. https://doi.org/10.3390/s21227692

Academic Editor: Steven Waslander

Received: 21 September 2021
Accepted: 17 November 2021
Published: 19 November 2021

Publisher's Note: MDPI stays neutral with regard to jurisdictional claims in published maps and institutional affiliations.

Copyright: © 2021 by the authors. Licensee MDPI, Basel, Switzerland. This article is an open access article distributed under the terms and conditions of the Creative Commons Attribution (CC BY) license (https://creativecommons.org/licenses/by/4.0/).

1. Introduction

Enabling robotised scanning processes is the harnessing of prior knowledge to fully traverse surfaces. For mobile or static-base robots completing NDT scans, knowledge of positions that have not been scanned is essential to ensure completeness of an inspection process that guarantees component integrity. Currently, this is ensured by planning a path over a known surface or part, that is then either verified of modified by an operator to ensure completeness.

Paths for parts equipped with an accurate CAD model can be produced automatically with commercial software. For parts without an accurate digital-twin, such as legacy parts or components with moulding errors, an operator has had to define a path on the robot's teach-pendant manually to capture its unique profile.

For one-off scans or for scanning parts with unique moulding errors, this process voids the high speed and repeatability benefits available to robotised NDT. In these cases, robotic platforms must be able to flexibly scan parts through online path planning, and to provide the same guarantee of completeness in surface coverage that is achieved by a human operator manually inspecting the part.

Recently, NDT has been enabled to define a 2-scan process. The first scan reconstructs the part for path planning of a subsequent scan with NDT equipment. The second scan can then commence, fully covering the known surface that is within reach of the robot. Methods of reconstructing part surfaces in the initial scan have been widely researched with respect to both Photogrammetry and in the field of machining.

In the field of Photogrammetry, automated robotised methods for free-form surface profiling have developed significantly. Processes involving 3D or 2D cameras have evolved from requiring user-inputted positions [1] to fully automated 3D model reconstruction techniques. Automated photogrammetry has been applied to a wide range of scales, from fine-detail model reconstruction using robotic arms [2,3] to large-scale reconstruction using autonomous robots with wide-aperture sensors [4]. A recent example of photogrammetry enabling a 2-pass scan within NDT utilising Structure-from-Motion (SfM) [5].

These methods have relied on multiple volumetric inspections of a complex object using wide field-of-view sensors such as traditional RGB or RGB + Depth (RGB/D) cameras. This work has considered surface profiling in the case of limited-range sensors, such as line-scanners or ultrasonic devices that have a field of view many magnitudes smaller than the inspected surfaces. In the case of laser scanners, a volumetric pre-scan is not safe for human operators working nearby. Volumetric scanning of curved objects cannot guarantee surface discovery in the case of water-coupled ultrasound devices without lengthy re-scanning processes due to beam divergence and scattering.

Within the field of machining, validation of machining quality or accurate part profiling when there is no available CAD model has been implemented using Coordinate Measuring Machines (CMMs). CMMs utilising limited field-of-view sensors for full-surface profiling have also been thoroughly investigated [6]. Their use has relied on spline-surface approximations to predict surface positions [7–9], or planar raster-tangent path planning [10]. These methods all require saturation of user-sampled positions, user input to define surface tangents, or rely on tangents defined by a gantry constrained rasterization pattern. The spline-surface approximation method has been successfully applied to ultrasonic-sensor surface discovery [11]. This method requires that the surface can be defined by a global spline, as opposed to an atlas of piece-wise smooth splines. This is disadvantaged when inspecting objects with discontinuities such as holes, as these cannot be captured by a global b-spline representation. Surfaces with global b-spline representations are also known as doubly ruled surfaces.

In aiding accurate offline path planning for Eddy-Current inspections, CMM machinery and software were applied within a manual pre-scan procedure to generate a CAD model [12].

This work has sought to completely remove the reliance on operator inputted information regarding the target surface, except for its maximal curvature. The authors have further aimed to completely automate the surface-profiling process, unconstrained by sensor type, robotic platform, or spline representations of the surface. The only requirement on sensor information is that the position of the surface relative to the sensor and the normal-direction of the surface are recoverable at each scan position. Approximate normal direction extraction requires discovery of at least 3 accurate local surface points.

Enabling full surface discovery requires a search process and memory structure to discover and store potential surface points for later traversal.

A candidate heuristic process are Flood Fill Algorithms (FFAs) that propagate through maps or networks in order to discover all positions within a connected surface or graph. The pseudo-code for two dimensional pixel maps has been presented in Algorithm 1 and accompanied by Figure 1.

Algorithm 1 Flood Fill algorithm on the plane.

1: FFA on the plane
2: Begin at Pixel P_1
3: $Open\text{-}List = \{P_1\}$;
4: $Points\text{-}Found = \{\}$;
5: **while** $|Open\text{-}List| > 0$: **do**
6:     $P_a = Open\text{-}List.back()$
7:     $Points\text{-}Found.insert(P_a)$;
8:     $Open\text{-}List.delete(P_a)$;
9:     **for** $direction \in \{'UP','DOWN','LEFT','RIGHT'\}$ **do**
10:        $P_b = P_a + direction$
11:        **if** P_b is new point AND not boundary point **then**
12:           $Open\text{-}List.insert(P_b)$;
13:        **end if**
14:     **end for**
15: **end while**

First iteration   Last iteration

Green: Found-points
Blue: Open-list
Black: Boundary points

Figure 1. Colour Flood-Fill on the plane.

This work has generalised planar FFA heuristics to three-dimensional surface traversal, inventing the Complete-Surface Finding Algorithm (CSFA). Whereas FFAs require a pre-known data structure, the novel CSFA requires only curvature information about the target surface to ensure complete coverage when applied to sensors of arbitrary dimensions and sensitivity.

Simple stack-based FFA and scanline heuristics are of particular interest in the simulation section. Scanline implementations choose a preferred direction of motion for search until a boundary position is reached. When a boundary position is discovered, the less-preferable step is then taken until a free path is found in the preferred direction of motion. The resultant path is a traditional rasterization pattern, which is widely utilised within NDT path planning operations.

FFAs have been applied in various contexts, due to their simplicity and versatility. In the context of image processing, FFAs have seen ongoing widespread use in commercial products as a time-efficient method for filling a bounded region with a given colour [13]. The principle of the bucket-fill programme has been inverted to aid segmentation algorithms in 2D and 3D contexts from a user-inputted mask [14–16]. In recent years FFAs have aided machine-learning programmes in object recognition through automatic mask generation [17]. Mixed mapping and network theoretic implementations have been implemented to guide image reconstruction. First, FFAs were shown to be as effective as quality guided algorithms [18], and subsequently used to enhance nearest neighbour node quality optimisation methods in various fields [19–21].

Further, FFA variants have been extensively implemented in robotic path planning and control. Discretised potential field variants such as modified CFill and Flood-Field Methods (FFMs) have been shown to have greater time efficiency in comparison to Potential Field Methods (PFMs) [22,23]. FFAs have gained interest in the context of optimal path planning for 2D platforms [24,25], that has demonstrated flexibility through effective integration with optimal motion planners such as the A* algorithm [26]. These concepts have evolved in application to optimal motion planning in 3D space for UAVs with an exhaustive search pattern [27]. Further FFA integration and heuristic mirroring has shown to enhance traditional path planning algorithms [28,29]. The above Flood-Fill methods have been implemented on data either with a pre-defined link structure or with a full exploration in each potential direction. For unknown surface profiling constrained by costly rearrangement procedures and a limited field of view, these procedures are either non-applicable or significantly sub-optimal.

2. Method

The aim of this paper has been to generate a complete set of points that describe the full surface by utilising the simple operations presented in Algorithm 1. To embed planar FFA operations within a 3D context requires the local position and normal direction information at each position.

A point source has been placed with a given stand-off from the surface in the normal direction, and a ray is then generated to intersect with the surface from which the tangent directions have been extracted. The 3D analogue of moving in the 2D principle directions is given by approximating the local surface covered by the sensor array with a tangent plane, defined by the observed points and approximate normal direction. Given a surface normal, the principal axes corresponding to 'UP' and 'DOWN' directions have been calculated through the Gram–Schmidt orthonormalization process [30]. Given a normal vector $\vec{n} = [n_x, n_y, n_z] = [n_i]$, and principle directions $\vec{e}_1 = [1,0,0]$, $\vec{e}_2 = [0,1,0]$ and $\vec{e}_3 = [0,0,1]$, the smallest component $\vec{x}$ has been selected as basis direction;

$$\vec{x} = \{\vec{e}_i \text{ if } |n[i]| = \min_{k \in [1,2,3]} |n[k]|\}. \tag{1}$$

The chosen basis direction has then been orthonormalised with the surface normal through the Gram–Schmidt process. The next basis direction $\vec{y}$ is taken by cross product of normal and tangent vectors. The basis directions $[\vec{x}, \vec{y}]$ have formed the cardinal directions that planar FFA's utilise of 'DOWN' and 'RIGHT'. The point source traverses the surface in an analogue implementation of the traditional planar FFA, displayed in Figure 2. If no data or insufficient data is available at a given position, the current search point is marked as being in the ambient space with no additional points hypothesised, representing the 3D analogue of a 2D boundary position.

The approximate local surface normal direction can be extracted from at least three distance measurements from a single position with a 2D sensor array. Well-calibrated 1D linear sensors arrays would require two measurement values within a small displacement range, and single-element 0D sensors would require data from at least three positions. The algorithm may be applied to any sensor capable of a surface-tool stand off measurement.

The authors have further adapted the simple embedded stack-based FFA implementation to produce a scanline variation that generates automatic rasterization patterns within post-processing. For surfaces with uni-directional curvature, this has been achieved by retaining the order of the extrapolated X, Y basis directions. Retaining order on surfaces with significant curvature in two directions, such as the sphere or bowl requires including a 'preferable direction' reference. This is so that when X and Y surface–tangent directions change their order during traversal, preference is given to the one that lies within a consistent plane in 3D space. On these surfaces, an irregular rasterization pattern emerges without preference vector. Irregular rasterization is not necessarily a negative feature, since for many robots and applications, there is an axial movement limit imposed that prevents multiple circular passes. This has been demonstrated in the results section, while rasterization is achieved in post-processing, online searches will require additional search positions that do not observe the target object in order to define boundary positions.

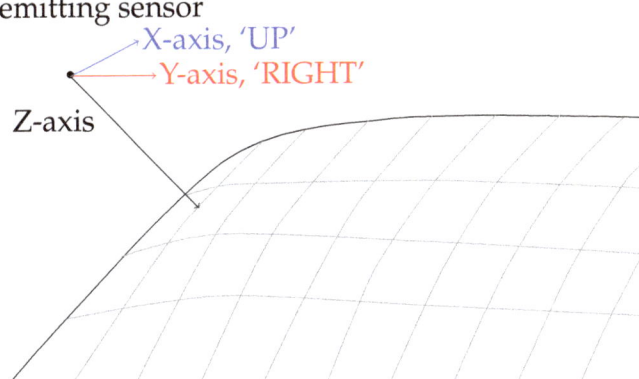

Figure 2. Flood Fill analogue in three dimensions. Grey lines represent iso-lines on the surface.

Finally a continuous surface must be discretised to ensure program closure, requiring a 3D analogue to 2D pixels. This structure allows positions that have been checked to be logged as seen. An Octree structure composed as a collection of boxes, or leaves has been chosen as it is less susceptible to numerical point-collisions present with a hash-table structure [31].

In order to assure full surface discovery, it is required that a step determined by the local information moves to a different Octree-node on the surface. Movements in 3D space under a set of changing basis directions may not align to a granular space oriented to the standard X, Y, Z bases. The undesirable effect of stepping within the same leaf may be prevented by moderating the Octree-leaf widths relative to the operator-specified step size d.

To ensure that each step defines a new leaf, the maximum potential length step within a leaf must be less than or equal to the step size. For leaf width w and step size d, the maximum step size, along the leaf's diagonal can be restricted with Equation (2).

$$w \leq \frac{d}{\sqrt{3}}. \tag{2}$$

On high-curvature surface sections the surface will inflect within each Octree leaf, reducing the Cartesian arc-length from one observed position to another. An upper bound for the arc-length reduction for curved surfaces needs to be defined to ensure that each step along the surface defines a new leaf.

Arc-length reduction due to the projection of a line along a curved surface is bounded by the surface's curvature, which defines how a local linearisation deviates from the true surface profile. This term has been defined for a small step-vector $\vec{dx}$ by the Second Fundamental Form (SFF) denoted II [32];

$$\text{Arc-length difference} \approx \vec{dx}^T \, \text{II} \, \vec{dx}/2. \tag{3}$$

The principal curvatures of the surface are eigenvalues of the SFF, and so the maximum possible inflection of a curve bound to the surface is in the direction of maximum principal curvature.

If the maximal principle curvature over the surface is κ_{max}, then an upper bound on the minimal required leaf-width for a step size d may be derived;

$$w/d \leq \frac{|1 - |\kappa_{max}|d/2|}{\sqrt{3}}. \qquad (4)$$

Dynamic discrete sampling may apply this principle to calculate minimal necessary Octree leaf-widths and step sizes in highly curved regions [33]; however, in this paper we restrict the analysis to uniform leaf-widths.

Flat surfaces have an over-sampling value of $w = d/\sqrt{3}$ (in units of d), since the maximal principal curvature is 0. This has returned Equation (2), since the step-size in ambient space is equivalent to that of the surface projection, the step taken always contained within the same spatial plane. An example of detrimental point-aliasing when curvature is not considered has been presented in the results section.

Finally, in the case of surfaces with a significantly restricted width, the step size should be limited to less than half of the minimum surface width.

The complete algorithm when simultaneously considering a pulse-echo test has been described in pseudo-code in Algorithm 2.

Algorithm 2 Pseudo-code for the novel CSFA.

1: Input: Maximum expected curvature κ, step-size d, and maximum Cartesian reach ΔX,
2: Octree = GenerateWorkSpace($\kappa, d, \Delta X$),
3: Operator moves sensor to surface,
4: GetData() →surface position and normal vector P_1, N_1,
5: Open-List = $\{P_1\}$
6: Points-Found = $\{\}$
7: **while** $|Open\text{-}List| > 0$ **do**
8:     $P_a = Open\text{-}List.back()$
9:     $Open\text{-}List.delete(P_a)$
10:     **if** $0 < |J_a^\Omega\{= InverseKin(P_a)\}|$ **then**
11:         Move to $J_a = \min_{motion} J_a^\Omega$
12:         GetData() → P_a, N_a, data
13:         **if** !data.$empty()$ **then**
14:             Sensor.$z_{direction}$ → N_a,
15:             GetUTdata(),
16:             Octree.insert(P_a)
17:             GramSchmidt(N_a) → {'UP','DOWN','LEFT','RIGHT'}
18:             **for** direction $\in$ {'UP','DOWN','LEFT','RIGHT'} **do**
19:                 $P_b = P_a + direction$
20:                 **if** $P_b \notin$ Octree **then**
21:                       Open-List.insert(P_b);
22:                 **end if**
23:             **end for**
24:         **end if**
25:     **end if**
26: **end while**

The CSFA process results in a single-pass process that reduces the overall number of steps, displayed in Figure 3.

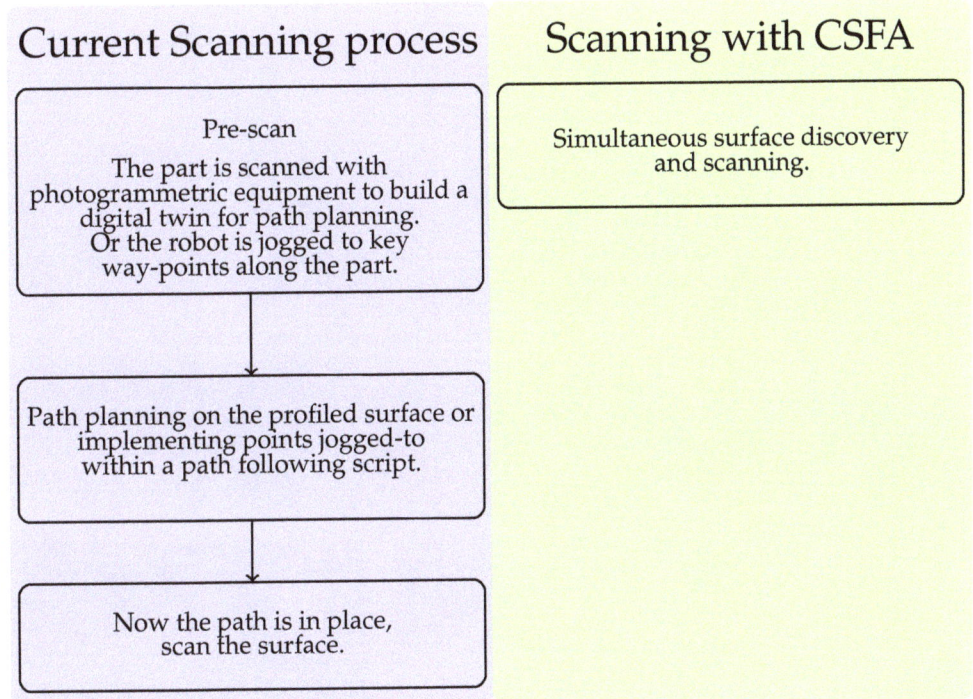

Figure 3. The one-step process enabled by the CSFA removes the necessity of accurate digital-twins and world-frame calibration, or lengthy robotic jogging procedures.

3. Robotic Path Planning

For robotic arm platforms, sections of the surface may lie out of reach, or a given motion may be impossible to execute due to a kinematic singularity [34]. These issues are incurred by a break in the correspondence between Cartesian space and the robot's fundamental coordinates, the possible joint-positions and link structure. In overcoming the spatial limitations of the robotic manipulator, oriented target-points were converted to configuration space coordinates. As a proof of concept investigation for the deployment of the novel CSFA, test pieces were chosen to test the algorithm's ability to ensure full coverage on curved and complex surfaces while minimising the risk of collision. Collision avoidance in the test cases were achieved by placing a motion-length limit. To maintain full coverage in the case of required back-tracking, any motion above this joint-space limit would cause the robot to move safely through a known point above the part. In the case of a convex part, point-to-point motion was considered admissible within one step if the subsequent point did not require motion in the current point's normal direction of more than the sensor-surface stand off. Since the algorithm requires an initial position to be defined along the surface, an initial configuration is given at the start. The path-planner then proceeded to choose the next in Cartesian space, and selected the candidate robotic configuration with the smallest joint-motion. If the selected point induced a configuration motion larger than the allowed threshold, the point was pushed back into the Open-List and another chosen until a suitable point was found or only large-motions were possible. In the latter case, the point with the smallest joint-wise motion was chosen. The robot was then sent joint-wise position command motions, avoiding kinematic singularities and ensuring the reachability of target points.

4. Results

Tests on shapes with key non-linear aspects have demonstrated the method's total coverage of generalised locally differentiable surfaces. The shapes chosen have been selected on the basis of surface irregularities that present challenges to full scanning. Surfaces with cut-outs that are not captured by a global surface spline representation demonstrated the advantage of the algorithm in handling machined parts, or in piecewise spline produced parts. These are not handled by the nearest available algorithm. Additionally, curved and doubly-curved surfaces were chosen to validate the suitability of the linearisation approximation method. In this section, surfaces chosen demonstrate complete coverage of locally smooth parts and parts with cut-outs. By demonstrating on positive, negative and zero curvature surfaces individually, the iterative and non-recursive algorithm has been validated for all locally smooth and holed surfaces. The process has been implemented in C++, utilising Simon Perrault's Octree structure [35]. Robotic simulations have been generated using RoboDK software with the Universal-Robotics UR10e as a demonstrative platform, with mesh simulations presented in MeshLab.

The CSFA has demonstrated ease in generating raster-motions on aerofoil components with varying step-sizes, displayed in Figure 4. Due to the relative flatness of the surface, a raster pattern was achieved. For more curved surfaces, there will be over-sampling of the space.

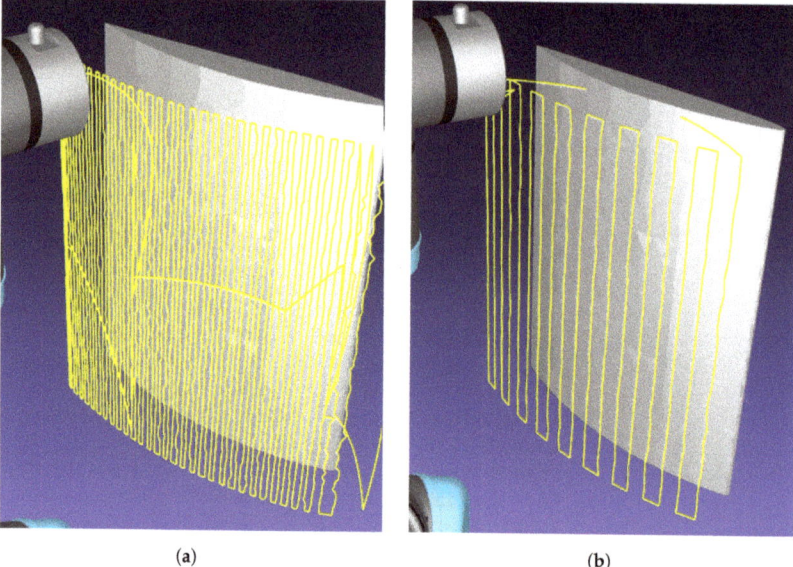

Figure 4. Demonstration of rasterizing a curved aerofoil component. The robotic path is traced in yellow, demonstrating the raster-like path obtained. (**a**) Sampling distance: 3 mm. (**b**) Sampling distance: 10 mm.

The method has been demonstrated to avoid surface-holes, re-scanning areas previously uncaptured in early-scan stages, displayed in Figure 5. The stack based memory of positions to check allowed effective full-surface discovery in the presence of irregular geometries. Figure 5 demonstrates that the CSFA has a clear advantage over gantry-based delivery platforms, covering complex surfaces without visiting the holed regions while still capturing the whole surface without needing the planar limits of the plate as input.

Repeatedly holed surfaces present multiple points of return, demonstrated in Figure 6.

The CSFA process makes a linear approximation of the surface in the neighbourhoods of discrete points. Displaying the algorithm on surfaces of positive and negative curvature, as in the sphere and bowl, demonstrates that it is robust in cases of local non-flatness. This is displayed in Figure 7.

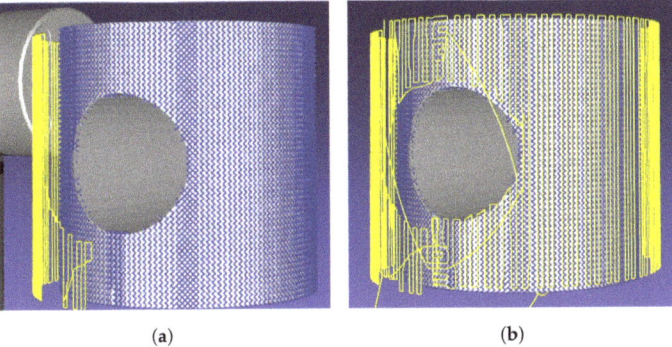

Figure 5. The scan initially misses sections of the pipe due to the shape's cross-sectional hole. The missed points are picked up at the end of the scan as there is memory of surface-positions to check. Points found are marked in blue, the robotic path traced in yellow. (**a**) Initial scan-pass. (**b**) End-of-scan.

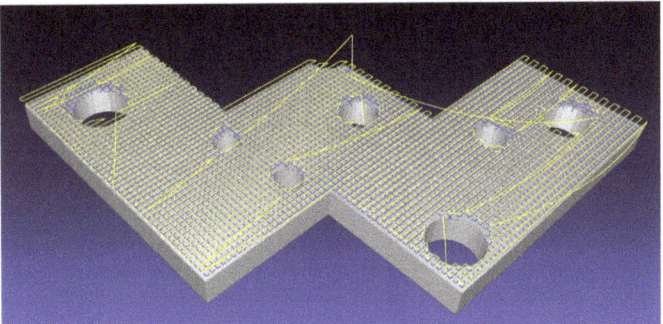

Figure 6. A complex flat plate holed with differently sized voids. The robotic path in yellow backtracks to allow for full surface discovery, shown by blue crosses, in the presence of surface-discontinuities.

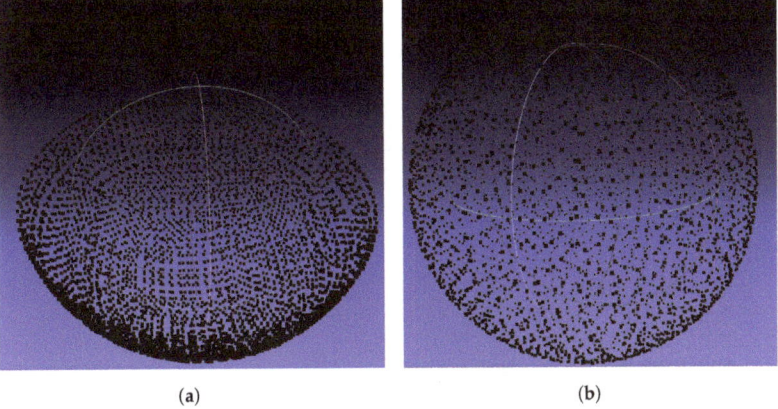

Figure 7. Points discovered while simulating a scan on a bowl and sphere of radius 150 mm with a sampling distance of 3 mm. (**a**) Concave shape sampling. (**b**) Sphere sampling.

The irregular rasterization pattern may be seen in Figure 8. Unlike for surfaces of only one direction of curvature such as in Figure 4 or Figure 5, rasterization for double-curvature surfaces is irregular. This incurs inefficient motions compared to traditional spiral-rasterization patterns.

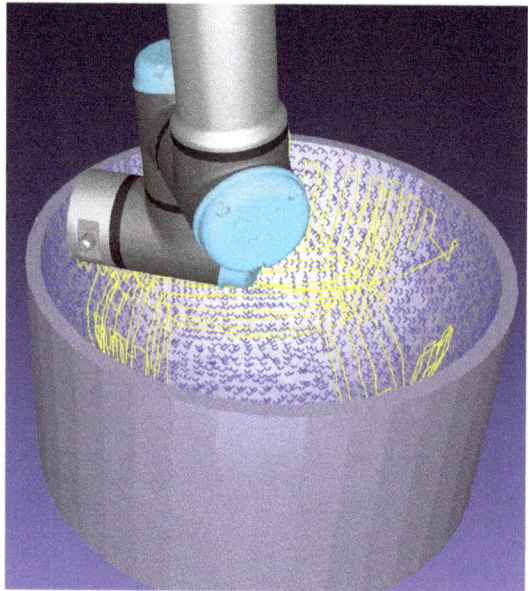

Figure 8. Sampling on a concave shape. The robotic path, that can form irregular patterns without a preferred direction, is shown in yellow. Discovered points on the bowl are shown as blue crosses.

A horizontal rasterization pattern of subsequent circles resembling traditional spiralized patterns may be imposed by using a preferred direction vector; however, they can result in large re-arrangement procedures seen in Figure 9.

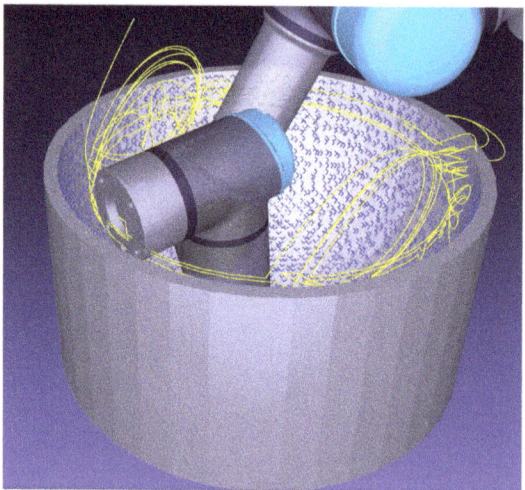

Figure 9. Sub optimal horizontal rasterization of a concave surface. Yellow trace lines demonstrate costly re-arrangement procedures to discover all the points shown in blue.

Curvature considerations are also demonstrably necessary for full surface coverage of components. Without over-sampling the space based on known surface curvature, full coverage is not guaranteed since taking a step will not necessarily take the algorithm to a new Octree-leaf. In turn, the algorithm stops prematurely as it aliases the points before and after the step within the Octree map. The effect of this is displayed in Figure 10.

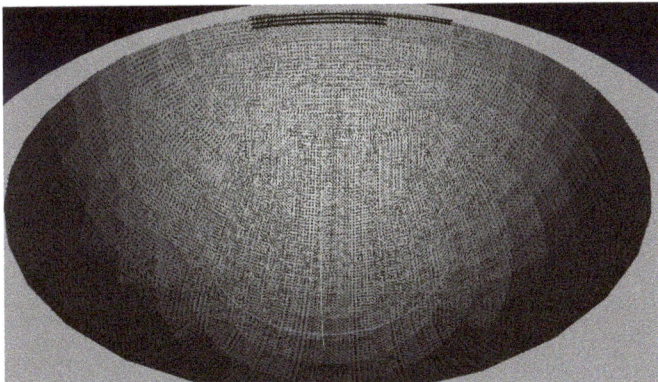

Figure 10. Points in bold display the extent of discovery with no over-sampling regime. Sampling rate: 1 mm, radius of bowl: 150 mm.

5. Experimental Results

Complete coverage of locally differentiable surfaces has been shown in simulation when there are no limitations due to the robotic platform or sensor. Two key test pieces were identified to validate the algorithm's practicality in deployment. These were a surface of doubled-curvature and a surface with a cut-out. The doubly curved surface has been chosen to show that with the correct step size, sensors with small ranges may complete the search process, and that the approximation found for the surface normal is a suitable one. Moreover, since the important quantity in Octree sampling to guarantee completeness is the ratio of curvature to step size, the doubly curved surface shows that the heuristic presented is applicable to surfaces of all curvatures, given a step size that does not hinder sensor-surface coupling. The part with a section cut out further validates the approach when the surface is not globally represented by a global b-spline, as is necessary within the nearest algorithm. Since the algorithm utilises an iterative and non recursive heuristic, by demonstrating the process on these surfaces it is also demonstrated to work on curved surfaces with varying curvature and with cut-outs. It is important to note that the hardware chosen for completing the scanning process is the limiting factor, as smaller sensors are necessary to complete scans on objects that have extreme curvatures.

Experimental testing of the CSFA utilised three flange-mounted Panasonic HG-C1030-P lasers, connected to an Arduino board for real-time data collection. The laser's viewing range was 30 mm ± 5 mm, limiting the feasible step size over highly curved surfaces, as height variations of over 5 mm over the step would remove the possibility of further surface discovery. The laser's repeatability did not affect motion planning, as it was in the range of 10 μm. The lasers were held within a 3D-printed cradle displayed in Figure 11. An external laptop collected data from the Arduino and Universal-Robots UR10e robotic platform simultaneously. Connecting through a COM port and Ethernet-enabled TCP/IP connection, respectively, position data and commands were received and sent to the robot. The CSFA, data interpretation, and inverse kinematics solutions were coded in C++. The external laptop had an Intel Core i5 processor with the program built and run from a Visual Studio programming environment. Results were imaged using Meshlab.

To represent a non globally smooth b-splineable surface, laminate plates were placed into a planar pattern with a cut out displayed in Figure 12a alongside the point-cloud of collected data displayed in Figure 12b. Full discovery of the target surface demonstrates the applicability of the CSFA in cases where a direct path along the surface to every point is not possible. The recollection of hypothesised points to visit allows traversal around corners, completely scanning regions with no direct path to one another.

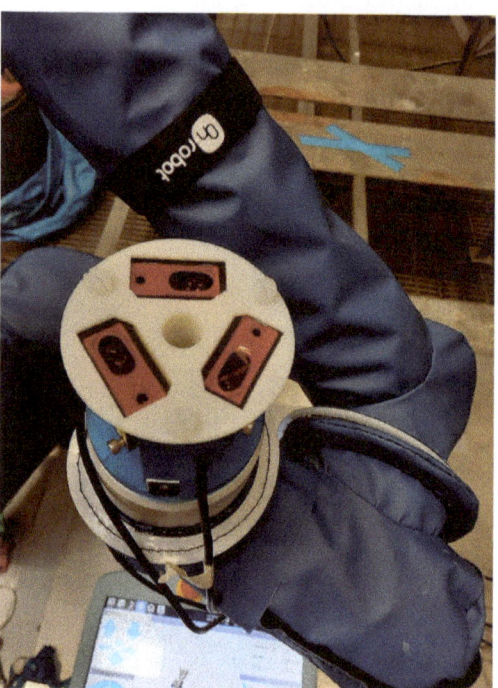

Figure 11. The tri-laser holder, attached to the UR10e flange. The design with rotational symmetry around axis 6 of the robot minimised the footprint of the tool.

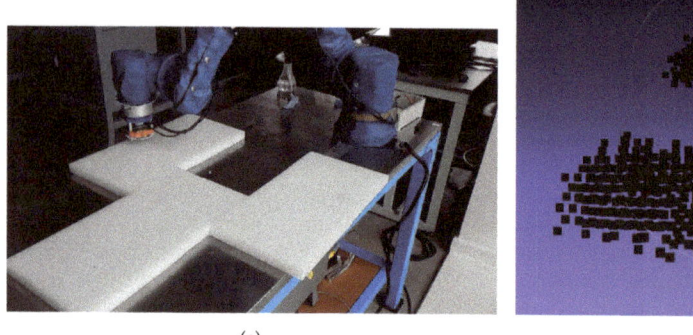

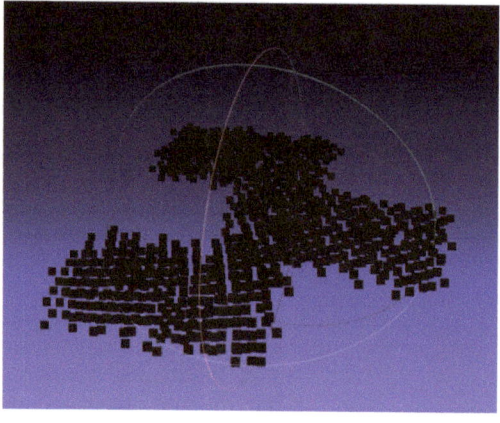

(a)   (b)

Figure 12. Automatic online profiling and scanning of an object with non-smooth shape. After a new point is found, the UT probe is applied to collect data. (**a**) Non-smooth shape created from arranged plates. (**b**) Resultant point cloud collected by the tri-laser and projected to the World-Frame using the live Joint-position of the robot.

A curved mock-aerofoil segment provided additional experimental data displaying application to a use-case commonly seen within NDT in Figure 13. The total time taken for this use-case was 7 min 30 s for 3 cm spaced collection points. Providing a real-world use-case for NDT, the full surface discovery of a doubly-curved surface with no-prior path planning provides the proof of concept for single-pass profiling of a complex surface and validation for the linearised surface approximation, while the part is relatively small compared to the robot's reach, the strength of this example is in the surface's extreme curvature. This use-case validates the application to surfaces commonly seen as complex within NDT.

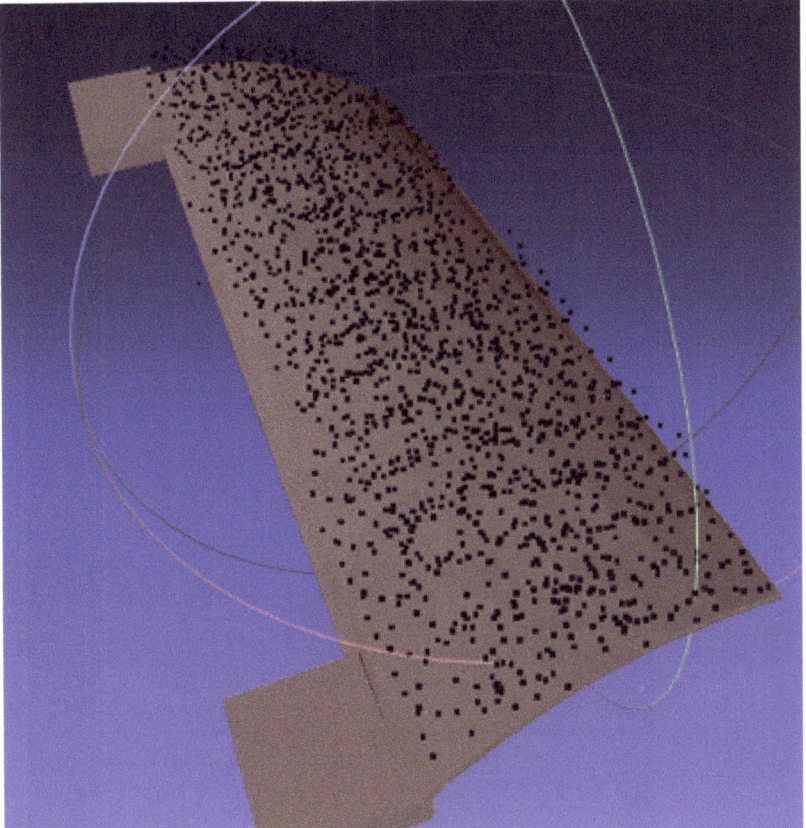

Figure 13. Point Cloud of a complex doubly-curved surface profiled in real time, aligned to the CAD model in post-processing.

Finally, the proof of concept for simultaneous non-contact surface profiling with the tri-laser platform combined with Conformable-Wedge-Probe scanning is presented. The process is two-step; the tri-laser discovers the surface, displayed in Figure 14a, the tool reversed and the Conformable Wedge Probe applied to the discovered position, displayed in Figure 14b.

In deployment, sensor ranges provided the most significant challenge. Since the tool's base had a diameter of 5 cm, the curvature of parts observed within that region had to not exceed the viewing range of the laser-sensors in order to ensure the tool and part did not collide.

The main source of risk to deployment was an incorrect laser-tool calibration. During early testing, the sensor's beam had an orientation offset that with larger step-sizes often risking collisions with the part. Scanning the planar part with a re-printed tool that corrected the laser-flange alignment, and calibrated using the four-point method, the standard deviation of points from the horizontal plane was 0.81 mm with mean signed-error of $O(10^{-16} \text{ mm})$.

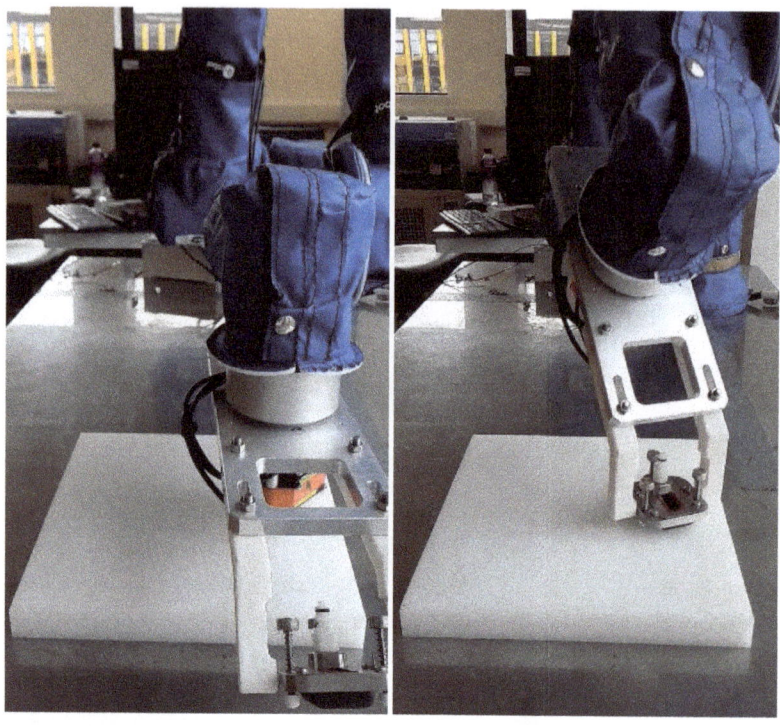

(a)   (b)

Figure 14. Automatic online single-pass profiling of a surface. (**a**) Initial non-contact surface discovery and profiling with the tri-laser. (**b**) Subsequent application of the Conformable-Wedge coupled UT device.

Further, while demonstrations were limited by the lack of a collision avoidance schema, these experiments have proven the algorithm's capability in autonomous scanning processes, and applicability to robotic NDT. The main challenge facing industrial deployment of robotic NDT where parts have no accurate digital-twin is the flexibility of the robotic platforms in use, and their ability to define complete surface coverage. We have proven the ability of this algorithm to overcome this issue in realistic contexts.

6. Discussion

The authors have successfully implemented an adaptation of the FFA for full coverage of free form surfaces. The implementation has been demonstrated on positive and negative curvature surfaces, highlighting how the linearised approximation is not a detriment to overall surface following capabilities of the algorithm.

In post-processing, the CSFA has been shown to output a raster-path along arbitrarily locally differentiable surfaces. For doubly-curved surfaces, the rasterization pattern becomes irregular and there is an over-sampling of points. However, the method ensures total coverage of the part which is preferable in NDT to sparse sampling. The potential applications of the algorithm are not limited to automatic rasterization procedures. The Octree memory method would allow fully automated discovery and scanning of structures with any robotic platform, such as mobile robots traversing a large structure. Further, the traversal method can be applied with any limited-aperture sensor, enabling a generalised surface-movement strategy when sensor data is limited. Finally, the discrete-point approach allows the method to capture surfaces that cannot be globally splined. The limitation in the case of significant surface discontinuities such as part-edges is that the process will not necessarily find the other side of the part, discovery determined by the sensor's range and aperture size relative to the discontinuity. In practical deployments the sensor range was the key limitation, limiting the sensors step size due to the surface curvature so as to continue full surface discovery. Practical demonstrations applied to complex cut-out surfaces and realistic doubly curved aerofoil mock-ups show the real-world application with limited-range laser sensors. Proof of concept for wedge-probe coupled UT applications provide the NDT specific aims of this paper of removing the need to path plan for full-surface scanning.

For complex surfaces such as aerofoils or machined plates with cut-outs, the algorithm demonstrated is safe for deployment. For more complex shapes such as external pipe-scans, limited knowledge of the environment is necessary to prevent collisions. Future work will deploy the algorithm using low-cost environmental sensors to prevent collisions and path planning such as Rapidly exploring Random Trees (RRT) algorithms to scan complex components.

Future works investigating online surface profiling will further consider options to remove the necessity for user-inputted curvature estimates and step-sizes entirely. Adaptations to specific sensor types for surface profiling shall also be considered.

Author Contributions: Conceptualization and Methodology, A.P. and M.S.; Investigation, Software, Validation, A.P.; Supervision, M.S., G.P. and A.G. All authors have read and agreed to the published version of the manuscript.

Funding: This research was funded by the Engineering and Physical Sciences Research Council (EPSRC) as part of an ICASE PhD studentship, grant number S513908/1.

Data Availability Statement: Not applicable.

Acknowledgments: Special thanks to David Carswell for code used within certain applications.

Conflicts of Interest: The authors declare no conflict of interest.

References

1. Callieri, M.; Fasano, A.; Impoco, G.; Cignoni, P.; Scopigno, R.; Parrini, G.; Biagini, G. RoboScan: An automatic system for accurate and unattended 3D scanning. In Proceedings of the 2nd International Symposium on 3D Data Processing, Visualization and Transmission, Thessaloniki, Greece, 6–9 September 2004.
2. Kriegel, S.; Rink, C.; Bodenmüller, T.; Suppa, M. Efficient next-best-scan planning for autonomous 3D surface reconstruction of unknown objects. *J. Real-Time Image Process.* **2013**, *10*, 611–631. [CrossRef]
3. Uyanik, C.; Secil, S.; Ozkan, M.; Dutagaci, H.; Turgut, K.; Parlaktuna, O. SPGS: A New Method for Autonomous 3D Reconstruction of Unknown Objects by an Industrial Robot. In *Towards Autonomous Robotic Systems*; Giuliani, M., Assaf, T., Giannaccini, M.E., Eds.; Springer International Publishing: Cham, Switzerland, 2018; pp. 15–27.
4. Almadhoun, R.; Abduldayem, A.; Taha, T.; Seneviratne, L.; Zweiri, Y. Guided Next Best View for 3D Reconstruction of Large Complex Structures. *Remote Sens.* **2019**, *11*, 2440. [CrossRef]
5. Khan, A.; Mineo, C.; Dobie, G.; Macleod, C.; Pierce, G. Vision guided robotic inspection for parts in manufacturing and remanufacturing industry. *J. Remanuf.* **2021**, *11*, 49–70. [CrossRef]
6. Kopáčik, A.; Erdélyi, J.; Kyrinovič, P. Coordinate Measuring Systems and Machines. In *Engineering Surveys for Industry*; Springer International Publishing: Cham, Switzerland, 2020; pp. 121–141. [CrossRef]
7. Lu, K.; Wang, W.; Wu, Y.; Wei, Y.; Chen, Z. An Adaptive Sampling Approach for Digitizing Unknown Free-form Surfaces based on Advanced Path Detecting. *Procedia CIRP* **2013**, *10*, 216–223. [CrossRef]
8. He, X.M.; He, J.F.; Wu, M.P.; Zhang, R.; Ji, X.G. Reverse Engineering of Free-Form Surface Based on the Closed-Loop Theory. *Sci. World J.* **2015**, *2015*, 903624. [CrossRef] [PubMed]
9. Zhang, Y.; Chen, K.; Guo, P.; Li, F.; Zhu, J.; Zhu, L.M. Profile tracking for multi-axis ultrasonic inspection of model-unknown free-form surfaces based on energy concentration. *Measurement* **2021**, *172*, 108867. [CrossRef]
10. Zhou, Z.; Zhang, Y.; Tang, K. Sweep scan path planning for efficient freeform surface inspection on five-axis CMM. *Comput.-Aided Des.* **2016**, *77*, 1–17. [CrossRef]
11. Guo, D.; Jiang, G.; Wu, Y.; Cheng, J. Automatic Ultrasonic Testing for Components with Complex Surfaces. In *DEStech Transactions on Engineering and Technology Research*; DEStech Publishing Inc.: Lancaster, PA, USA, 2017.

12. Morozov, M.; Pierce, S.; MacLeod, C.; Mineo, C.; Summan, R. Off-line scan path planning for robotic NDT. *Measurement* **2018**, *122*, 284–290. [CrossRef]
13. MS Windows Developer FloodFill Function. Available online: https://docs.microsoft.com/en-us/windows/win32/api/wingdi/nf-wingdi-floodfill (accessed on 19 March 2021).
14. Bhargava, N.; Trivedi, P.; Toshniwal, A.; Swarnkar, H. Iterative Region Merging and Object Retrieval Method Using Mean Shift Segmentation and Flood Fill Algorithm. In Proceedings of the 2013 Third International Conference on Advances in Computing and Communications, Mumbai, India, 18–19 January 2013; pp. 157–160. [CrossRef]
15. Chu, P.; Cho, S.; Park, Y.; Cho, K. Fast point cloud segmentation based on flood-fill algorithm. In Proceedings of the 2017 IEEE International Conference on Multisensor Fusion and Integration for Intelligent Systems (MFI), Daegu, Korea, 16–18 November 2017; pp. 656–659.
16. Lee, T.; Lim, S.; Lee, S.; An, S.; Oh, S. Indoor mapping using planes extracted from noisy RGB-D sensors. In Proceedings of the 2012 IEEE/RSJ International Conference on Intelligent Robots and Systems, Vilamoura, Algarve, Portugal, 7–12 October 2012; pp. 1727–1733. [CrossRef]
17. He, Y.; Hu, T.; Zeng, D. Scan-flood Fill(SCAFF): An Efficient Automatic Precise Region Filling Algorithm for Complicated Regions. *arXiv* **2019**, arXiv:1906.03366.
18. Chen, K.; Xi, J.; Yu, Y. Fast quality-guided phase unwrapping algorithm for 3D profilometry based on object image edge detection. In Proceedings of the 2012 IEEE Computer Society Conference on Computer Vision and Pattern Recognition Workshops, Providence, RI, USA, 16–21 June 2012; pp. 64–69. [CrossRef]
19. Li, Y.; Cui, X.; Wang, H.; Zhao, M.; Ding, H. Comparison of phase unwrapping algorithms for topography reconstruction based on digital speckle pattern interferometry. In *AOPC 2017: Optical Spectroscopy and Imaging*; Society of Photo-Optical Instrumentation Engineers (SPIE) Conference Series; Yu, J., Wang, Z., Hang, W., Zhao, B., Hou, X., Xie, M., Shimura, T., Eds.; SPIE: Bellingham, DC, USA, 2017; Volume 10461, pp. 450–461. [CrossRef]
20. Li, Q.; Bao, C.; Zhao, J.; Jiang, Z. A New Fast Quality-Guided Flood-Fill Phase Unwrapping Algorithm. *J. Phys. Conf. Ser.* **2018**. *1069*, 012182. [CrossRef]
21. Chen, K.; Xi, J.; Yu, Y.; Chicharo, J. Fast quality-guided flood-fill phase unwrapping algorithm for three-dimensional fringe pattern profilometry. In *SPIE/COS Photonics Asia*; SPIE: Bellingham, DC, USA, 2010.
22. Zmudzinski, L. Rough Mereology Based CFill Algorithm for Robotic Path Planning (short paper). In *Proceedings of the 28th International Workshop on Concurrency, Specification and Programming, Olsztyn, Poland, 24–26 September 2019*; Ropiak, K., Polkowski, L., Artiemjew, P., Eds.; CEUR Workshop Proceedings; CEUR-WS.org; University of Warmia and Mazury: Olsztyn, Poland, 2019; Volume 2571.
23. Elshamarka, I.; Saman, A. Design and Implementation of a Robot for Maze-Solving using Flood-Fill Algorithm. *Int. J. Comput. Appl.* **2012**, *56*, 8–13. [CrossRef]
24. Kibler, S.; Raskovic, D. Coordinated multi-robot exploration of a building for search and rescue situations. In Proceedings of the 2012 44th Southeastern Symposium on System Theory (SSST), Jacksonville, FL, USA, 11–13 March 2012; pp. 159–163. [CrossRef]
25. Tjiharjadi, S.; Setiawan, E. Design and Implementation of a Path Finding Robot Using Flood Fill Algorithm. *Int. J. Mech. Eng. Robot. Res.* **2016**, *5*, 180–185. [CrossRef]
26. Tjiharjadi, S.; Wijaya, M.; Setiawan, E. Optimization Maze Robot Using A* and Flood Fill Algorithm. *Int. J. Mech. Eng. Robot. Res.* **2017**, *6*, 366–372. [CrossRef]
27. Ranade, S.; Manivannan, P.V. Quadcopter Obstacle Avoidance and Path Planning Using Flood Fill Method. In Proceedings of the 2019 2nd International Conference on Intelligent Autonomous Systems (ICoIAS), Singapore, Singapore, 28 February–2 March 2019; pp. 166–170. [CrossRef]
28. Kalisiak, M.; van de Panne, M. RRT-blossom: RRT with a local flood-fill behavior. In Proceedings of the ICRA 2006 IEEE International Conference on Robotics and Automation, Orlando, FL, USA, 15–19 May 2006; pp. 1237–1242. [CrossRef]
29. Guo, J.; Lin, Y.; Su, K.; Li, B. Motion Planning of Multiple Pattern Formation for Mobile Robots. *Appl. Mech. Mater.* **2013**, *284–287*, 1877–1882. [CrossRef]
30. Cheney, W.; Kincaid, D. *Linear Algebra: Theory and Applications*; G—Reference, Information and Interdisciplinary Subjects Series; Jones & Bartlett Learning: Burlington, MA, USA, 2012.
31. Meagher, D. *Octree Generation: Analysis and Manipulation*; Defense Technical Information Center: Fort Belvoir, VA, USA, 1982.
32. Chase, H. Fundamental Forms of Surfaces and the Gauss-Bonnet Theorem. 2012. Available online: https://math.uchicago.edu/~may/REU2012/REUPapers/Chase.pdf (accessed on 1 September 2021).
33. Kotani, M.; Naito, H.; Omori, T. A discrete surface theory. *Comput. Aided Geom. Des.* **2017**, *58*, 24–54. [CrossRef]
34. Simaan, N.; Shoham, M. Singularity analysis of a class of composite serial in-parallel robots. *IEEE Trans. Robot. Autom.* **2001**, *17*, 301–311. [CrossRef]
35. Simon Perreault's C++ Octree Implementation. Available online: https://nomis80.org/code/octree.html (accessed on 19 March 2021).

Article

Automated Real-Time Eddy Current Array Inspection of Nuclear Assets

Euan Alexander Foster [1,*], Gary Bolton [2], Robert Bernard [3], Martin McInnes [1], Shaun McKnight [1], Ewan Nicolson [1], Charalampos Loukas [1], Momchil Vasilev [1], Dave Lines [1], Ehsan Mohseni [1], Anthony Gachagan [1], Gareth Pierce [1] and Charles N. Macleod [1]

1. SEARCH: Sensor Enabled Automation, Robotics & Control Hub, Centre for Ultrasonic Engineering (CUE), Department of Electronic & Electrical Engineering, University of Strathclyde, Royal College Building, 204 George Street, Glasgow G1 1XW, UK
2. National Nuclear Laboratory LTD., Warrington WA3 6AE, UK
3. Sellafield LTD., Sellafield, Seascale, Cumbria CA20 1PG, UK
* Correspondence: e.foster@strath.ac.uk

Abstract: Inspection of components with surface discontinuities is an area that volumetric Non-Destructive Testing (NDT) methods, such as ultrasonic and radiographic, struggle in detection and characterisation. This coupled with the industrial desire to detect surface-breaking defects of components at the point of manufacture and/or maintenance, to increase design lifetime and further embed sustainability in their business models, is driving the increased adoption of Eddy Current Testing (ECT). Moreover, as businesses move toward Industry 4.0, demand for robotic delivery of NDT has grown. In this work, the authors present the novel implementation and use of a flexible robotic cell to deliver an eddy current array to inspect stress corrosion cracking on a nuclear canister made from 1.4404 stainless steel. Three 180-degree scans at different heights on one side of the canister were performed, and the acquired impedance data were vertically stitched together to show the full extent of the cracking. Axial and transversal datasets, corresponding to the transmit/receive coil configurations of the array elements, were simultaneously acquired at transmission frequencies 250, 300, 400, and 450 kHz and allowed for the generation of several impedance C-scan images. The variation in the lift-off of the eddy current array was innovatively minimised through the use of a force–torque sensor, a padded flexible ECT array and a PI control system. Through the use of bespoke software, the impedance data were logged in real-time ($\leq$7 ms), displayed to the user, saved to a binary file, and flexibly post-processed via phase-rotation and mixing of the impedance data of different frequency and coil configuration channels. Phase rotation alone demonstrated an average increase in Signal to Noise Ratio (SNR) of 4.53 decibels across all datasets acquired, while a selective sum and average mixing technique was shown to increase the SNR by an average of 1.19 decibels. The results show how robotic delivery of eddy current arrays, and innovative post-processing, can allow for repeatable and flexible surface inspection, suitable for the challenges faced in many quality-focused industries.

Keywords: non-destructive evaluation; robotic NDE; automated eddy current testing; eddy current arrays

1. Introduction

The global Non-Destructive Testing (NDT) market size was valued at USD 6.3 billion in 2021 with a predicted compound annual growth rate (CAGR) of 13.66% from 2022–2029 to hit a total market value of USD 16.66 billion [1]. This high level of growth can be attributed to the rise of "NDT 4.0", in which greater connectivity across the manufacturing supply chain is sought through the integration of connected sensors of which NDT techniques play a role [2]. To deliver this level of interconnectivity, it is now commonplace to see automated robotic delivery of NDT [3–6].

The vast majority of the NDT market is based on volumetric inspection of high-value infrastructure and components, such as automotive/aerospace components or public rail infrastructure, primarily through the use of radiographic and ultrasonic testing. Due to this popularity, the automation of volumetric techniques is the most mature in the NDT industry. Further growth in the automation of volumetric NDT is expected to lag behind other NDT techniques, as innovation has shifted towards more novel and complex delivery of volumetric NDT as well as incorporating advanced imaging and post-processing techniques. Examples of these trends include performing the volumetric inspection at the point of manufacture for high-value components [7–12], performing aerial UAV-based volumetric inspection [13–16], optimising the amount of data gathered [17,18], and deploying machine/deep learning in the analysis of the datasets generated [19–21].

By contrast, the automation of surface inspection is far less mature and from 2022–2029 it is predicted to have the highest CAGR of any NDT technique due to the increased adoption of Eddy Current Testing (ECT) [1]. Of the 'big 5' NDT techniques, eddy current, magnetic particle, and penetrant testing were shown to be able to detect surface-breaking flaws, where others in the 'big 5' (ultrasound and radiographic) struggle [22].

Eddy currents are induced in a sample according to Faraday's Law of Induction [23] when a coil carrying an alternating current produces an alternating magnetic field and the conductive sample lies within this magnetic field. The induced eddy current in the sample is of the opposite phase to that of the coil and sets up its own magnetic field to oppose that of the coil. The eddy current density, $J(z)$, decays exponentially with depth z in an isotropic material, and the sensed impedance is directly proportional to the current density [24]:

$$J(z) = J_0 \exp\left(-\frac{z}{\delta_{sd}(1+i)}\right) \quad (1)$$

In the presence of a defect, the current density is altered and this change in current density can be sensed as a change in impedance. The magnitude of the eddy current density decays exponentially and when it falls to $1/e$ of its surface value, the depth at which this occurs is known as the standard depth of penetration, δ_{sd}. The standard depth of penetration is dependent on the frequency of the voltage in the coil, the magnetic permeability, and the electrical conductivity of the component, and is widely viewed as the deepest depth a meaningful change in impedance can be sensed. Due to the exponential decay associated with eddy currents, they are ideally suited to detecting surface-breaking defects. This is in direct contrast with ultrasound where the front wall echo typically masks any shallow defects within a component. With correct eddy current probe design and frequency selection, an eddy current can be created that has a standard depth of penetration greater than or equal to the thickness of some thin-walled components, such as the canisters used in the storage of low-level nuclear waste.

Magnetic particle testing is restricted to the use of ferromagnetic metals and requires the component to be magnetized/de-magnetized frequently. While penetrant testing is not restricted to any material but requires the component to be coated in a penetrant and developer, which is frequently undesirable. Both magnetic particle and penetrant testing are subject to great operator error and do not produce discrete data points as a sensor is rastered across the component's surface making automation unfeasible. However, these drawbacks do not exist for ECT, and hence ECT is well suited for automation. As society moves towards Industry 4.0, automation is becoming increasingly important in surface inspection in the immediate future.

In comparison to volumetric techniques, ECT does not suffer from the health and safety concerns associated with radiographic inspection. Additional technical requirements may also prohibit the use of other inspection modalities. For example, multi-angle accessibility requirements and part size limitations may make computed tomography radiographic testing unfeasible. While for ultrasonic inspection, environmental factors may deter the use of a couplant. ECT has a significant advantage as single-sided access is all that is required, and no couplant is needed to perform an inspection.

Reuse and sustainable business practices are the main drivers behind the increased adoption of ECT, as detecting surface-breaking flaws that occur in operation is becoming increasingly important to prolong the safe operation of key assets for industries such as nuclear and aerospace. Furthermore, due to the lower market size, robotic delivery of ECT is far less common with only a few primitive integration efforts being reported in the literature [25–28]. To keep pace with the high throughput demands of modern production/maintenance lines, increased robotic deployment of ECT is necessary and vital to capitalise on this demand.

This paper presents, for the first time, the automated deployment of an eddy current array, via a flexible robotic cell complete with force–torque control, to scan a canister typical of the ones used in the storage of spent nuclear fuel. Table 1 shows a comparison between previously published papers that feature robotically deployed eddy current inspection and this work. Real-time adaptive control of a 6-axis robotic arm (KUKA Quantec Extra HA KR-90 R3100, Augsburg, Bavaria, Germany [29]) and an external rotary stage (KUKA DPK-400 [30]) with force–torque compensation was accomplished using a framework described in the author's previous work [31]. Force–torque compensation allowed for constant lift-off of the eddy current array during the inspection. This was intentionally carried out as it was shown that robotically delivered eddy current inspection offers far less noise when compared to that of manual eddy current inspection [32]. A commercial 32-element padded eddy current array from EddyFi (Part No: ECA-PDD-034-500-032-N03S, Québec, QC, Canada [33]) with a centre frequency of 500 kHz and an operating frequency range of 100–800 kHz, along with a 64-element Eddyfi Ectane 2 controller [34] were used to perform 180-degree rotary scans of a 1.4404 stainless steel nuclear grade canister with known stress corrosion cracks. Extensive software infrastructure coupled with the Eddyfi Ectane 2 Software Development Kit (SDK) allowed for the impedance data to be logged and analysed in real-time. All data were stored in a proprietary binary file format to allow for further post-processing in MATLAB.

Table 1. Comparison between previously published robotically deployed eddy current inspections and this work.

	Mackenzie et al., 2009 [25]	Summan et al., 2016 [26]	Morozov et al., 2018 [27]	Zhang et al., 2020 [28]	This Work
Adaptive Motion	✗	✗	✗	✗	✓
Eddy Current Array	✗	✓	✗	✓	✓
Image Compensation	✗	✗	✗	✗	✓

Where ✓ denotes yes and ✗ denotes no.

This infrastructure allowed for the acquisition and real-time analysis of impedance data. Novel image post-processing techniques, such as phase rotation and mixing, were shown to increase the Signal to Noise Ratio (SNR) of the resulting C-scan images by an average of 4.56 and 1.19 decibels, respectively. It is envisaged that studies such as this will progress eddy current testing to match the level of flexibility and quality enjoyed in the post-processing of ultrasonic datasets [35,36].

2. Experimental System

NDT is crucial to safety-conscious industries such as nuclear [37]. Traditionally, the inspection of nuclear assets is highly resource-intensive and complex. The inspection challenge is complicated further when the asset lifetime exceeds the original design intent. This problem is one that is currently being faced in the UK, where government policy has shifted from favouring reprocessing to long-term storage of nuclear assets [38]. Spent nuclear fuel due for reprocessing is now being stored long term as reprocessing facilities are closed down. Some sites store low-level waste in canisters made from 0.9 mm thick 1.4404 stainless steel. These canisters range from 130–150 mm in diameter and are ~300 mm in length. To allow for effective cooling, the canisters are stored in facilities that are partially open to the environment. Given the coastal location of the UK, stress corrosion cracking is a concern. Due to the points mentioned above, canisters with intentionally induced stress corrosion cracks were scanned with an eddy current array and the acquired impedance data were analysed within this study.

2.1. Hardware and Experimental Summary

Figure 1 shows the experimental hardware used in the automated deployment of the eddy current system. A nuclear canister with a matrix of 16 stress corrosion cracks shown in Figure 2 is held within a mechanical chuck on top of a KUKA DPK-400 external rotary stage that has an angular resolution of 0.009°. The padded Eddyfi eddy current array (Part No: ECA-PDD-034-500-032-N03S) is mounted in a bespoke 3D-printed housing which is in turn secured to an IP-65 rated gamma force–torque sensor from ATI Industrial Automation (Apex, NC, USA) [39]. To move the sensor to the height of interest for the inspection, the eddy current array, 3D-printed housing and force–torque sensor assembly, are mounted to the flange of a KUKA KR-90 robot. Both the KR-90 and DPK-400 external rotary stages are controlled via a KRC 4 controller [40].

Figure 1. Eddy current inspection hardware.

In order for the eddy current array to be pressed onto the canister surface in the direction of the canister' centre, a calibration tool was manufactured to teach the KR-90 robot a new base coordinate system. The calibration tool was made so that it would align the centre of the chuck to the centre of the rotary stage. Additionally, the calibration tool allowed for the centre of the tool along with 4 concentric radial calibration points at 150 mm in 90° increments to be taught to the KR-90 robot. By teaching the KR-90 robot these points, it was able to know where the centre of the canister and rotary stage was relative to its own coordinate system, and ensure motion was performed relative to this point. This effort guaranteed that the eddy current array was always pressed against the canister surface in the direction of the canister's axial centre and helped establish good electromagnetic coupling during the automated inspection.

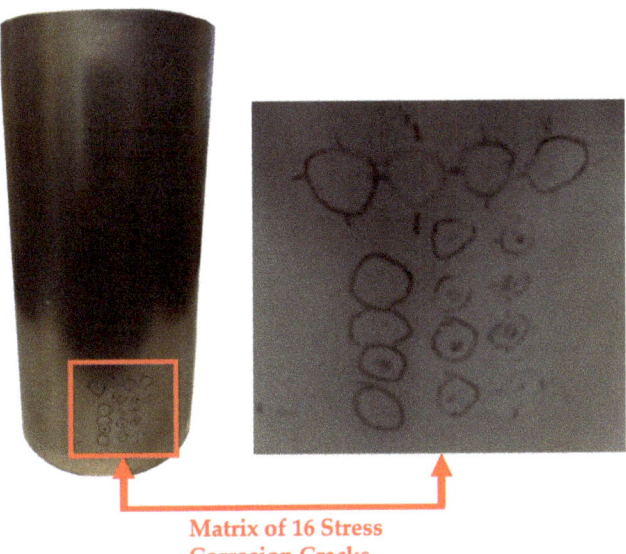

Figure 2. Canisters with a matrix of 16 stress corrosion cracks. Depositions of 5 µL droplets of sea water, 3.03 g/L of MgCl2, 15.2 g/L of MgCl2 and 30.03 g/L of MgCl2 were used to induce the cracks in the top row, left, central and right columns, respectively.

The eddy current array is deployed to the height of interest in the Z-direction of the canister via a variable set by the user on the Graphical User Interface (GUI) of a LabVIEW program using a framework similar to previously published work [31]. Force and torque in and around all three axes shown in Figure 1 are sensed via the force–torque sensor and are transmitted to a LabVIEW program via the robot controller using the Kuka Robot Sensor Interface (RSI) [41]. The transmission of the force and torque characteristics allowed for: (1) the adaptive motion of the eddy current sensor during inspection; (2) the balancing of the eddy current probe and the subsequent triggering for the acquisition of the impedance data to begin; and (3) the triggering of the rotary stage to begin movement. It is important to note that the force–torque sensor was calibrated with all hardware mounted prior to any automated inspection through a program provided by the manufacturer. The calibration enabled the net force and torque values being applied to the eddy current array and mounting assembly to be correctly sensed and subsequently transmitted to the LabVIEW control program for adaptive motion to be performed.

The KR-90 robot presented the eddy current array onto the surface of the nuclear canister at the user-specified height, and a target force and torque of 10 N and 0 Nm were met in the Y-direction and X-axis, respectively, for 3 s. Once this time period had passed, the balancing of all coils within the eddy current array was performed when the probe was stationary. After a further 3 s, the impedance data acquisition along with the rotary stage movement was triggered.

During the inspection, a PI control system was used to monitor and correct both the force in the Y-direction and the torque around the X-axis at the previously mentioned target force and torque values. It was found that *P*- and *I*-values of 0.1 and 0.0 gave an adequate control response. Control of the eddy current probe's orientation in this manner allowed for minimal variations in the lift-off of the eddy current array throughout the inspection providing excellent coupling. Other previously published literature has shown that lift-off can be reduced via advanced signal processing and elaborate probe design [42]. These efforts are often particularly involved and particular to one sample/defect type. As a result, these lift-off compensation strategies are complex to deploy and benefit from. The approach in this paper of utilising a force–torque sensor in combination with a padded ECT array provides experimental flexibility and passively compensates for any lift-off variation at the point of acquisition giving wide-reaching benefits.

The acquisition of the impedance data was stopped when the rotary stage had completed the angular movement requested by the user from within the LabVIEW program. A singular scan can be summarised by the following process:

1. A connection with the eddy current Ectane device is made.
2. The eddy current array is set up with the following parameters:
    a. Probe type;
    b. Probe configuration (axial and/or transversal—See Section 2.2);
    c. Frequencies;
    d. Voltages;
    e. Gain;
    f. Repetition rate.
3. The robot and external rotary axis are set up with the following parameters:
    a. Linear speed of the KR-90 robot;
    b. Approach speed of the KR-90 robot;
    c. Angular movement of the canister/external rotary stage;
    d. Angular speed of the canister/external rotary stage;
    e. Target force for the KR-90 robot to apply the array onto the canister.
4. The KR-90 robot places the probe against the canister and the target force is reached.
5. The target force is maintained for 3 s.
6. The balancing of the eddy current array is performed.
7. Wait a further 3 s.
8. The acquisition of impedance data and rotary stage movement is triggered.
9. Once full angular motion is complete, the acquisition of impedance data is stopped.
10. The KR-90 robot moves the eddy current array to a predetermined safe position.
11. The acquired impedance data are saved to a binary file for post-processing in MATLAB.

2.2. Eddy Current C-Scan Acquisition

Figure 3 shows a generic eddy current array layout along with illustrations of the transmit and receive pairings for the axial and transversal configurations. Depending on the probe geometry, there may or may not be an equal number of transversal and axial transmit and receive pairs. Each pairing in each configuration generates a data point of complex impedance data. The probe is linearly scanned perpendicular to the coil columns as noted in Figure 3, and the data points are logged into a complex 2D array. The resulting complex arrays can then be post-processed, and the vertical component of the post-processed complex array can be plotted in a C-scan format to show any defective signals with maximum Signal to Noise Ratio (SNR).

As can be seen in Figure 3b for the axial configuration, coils in the array are excited in one column and reception of the impedance data is performed across the array in the second column. Conversely, the transversal configuration documented in Figure 3c shows coils being excited and reception of the impedance data being performed within the same vertical column of coils.

The coil firing sequence is changed between the axial and transversal configurations to achieve greater sensitivity to differing defect orientations. With reference to the coordinate system in Figure 3, a larger change in impedance would be observed for a defect that is aligned with the X-axis for a transversal configuration over that of an axial configuration. This is due to the defect more severely intercepting the eddy current that exists between the two transmit and receive coils in the transversal configuration over that of the axial configuration. This greater compression of the eddy currents caused by the defect presence will have a large effect on the electromagnetic field and by proxy the sensed change in impedance. The opposite can be said to be true for a defect aligned in the Y-direction. For further reading, Ye et al. [42] provide a thorough theoretical and experimental investigation of this phenomenon.

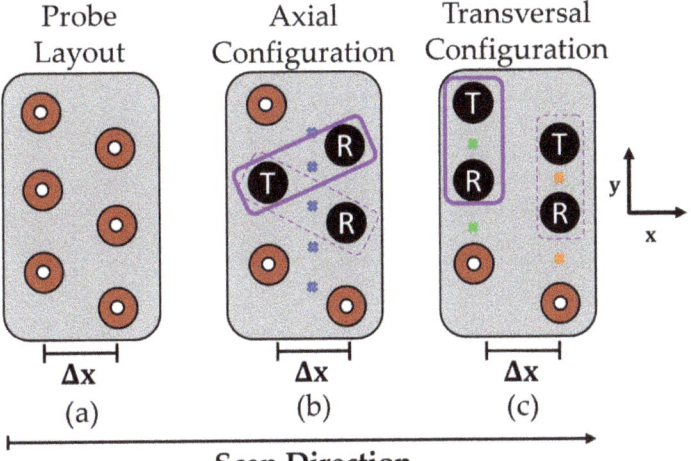

Figure 3. Eddy current array transmit and receive configurations. (**a**) Generic Eddy current array layout with two vertical columns of coils. (**b**) Axial transmit and receive configuration where x (in blue) corresponds to the transmit/receive pair centres of the excited eddy current channels in the test part. (**c**) Transversal transmit and receive configuration where x (in green) corresponds to the transmit/receive pair centres of the excited eddy currents in the test part resulting from the first/odd column of coils, and where x (in orange) corresponds to the transmit/receive pair centres of the excited eddy currents in the test part resulting from the second/even column of coils.

It is also evident from Figure 3 that the centres of excitation are not aligned between the axial and transversal datasets in the X-direction. Moreover, for each coil column within the transversal dataset, the data centres are also misaligned. As alluded to in Section 2.2, this positional misalignment is corrected within the LabVIEW program and ensures that the resulting complex array for each dataset has the same spatial grid.

Key to the positional compensation is the acquisition rate of the eddy current array and the angular speed of the rotary stage so that each acquisition point aligns with an integer number of divisions of half the array coil column pitch, Δx. The acquisition rate and number of divisions between half of the array column pitch are set by the user, and the coil pitch is defined by the geometry of the array. These three variables are used to set the angular speed of the rotary stage. For example, if an eddy current array has a column coil pitch of $\Delta x = 7$ mm, an acquisition rate of 50 Hz, and 50 divisions, the linear speed would need to be $\left(\frac{7}{2} \times \frac{1}{50}\right) / \left(\frac{1}{50}\right) = 3.5$ mm/s. This linear speed can then be converted to rotational speed by dividing the diameter of the canister at 150 mm to give the angular speed of the rotary stage at 1.34 deg/s. Whilst individual datapoints are not positionally-encoded, the positional location is extrapolated from setting the angular speed relative to the eddy current probe geometry and acquisition rate as mentioned above. By doing so, it ensures that data are acquired at both the axial and transversal data centre points on the X-axis as the array is linearly scanned.

In order to ensure a common spatial grid, the first and last impedance data points corresponding to a distance of half the coil pitch are discarded within the axial complex array. By discarding the first set of data points that cover half the coil pitch, the axial complex array in the X-direction is synched with the first/odd column of the transversal dataset. Moreover, by discarding the last set of data points that cover half the coil pitch, the axial complex array in the X-direction is synched with the second/even column of the transversal complex array. This discarding of data is shown graphically in Figure 4a. The resulting data is then linearly interpolated in the Y-direction to align with the Y-coordinates of the transversal complex array.

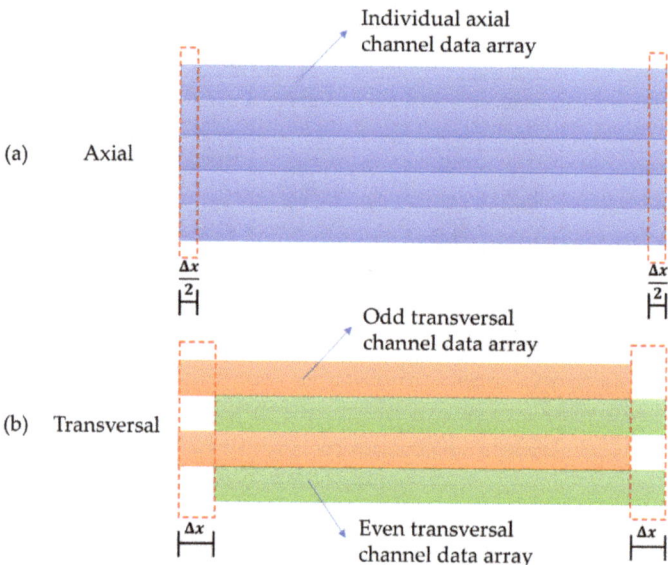

Figure 4. Illustration of complex impedance data positional compensation performed between axial and transversal configurations. (**a**) Axial complex array positional compensation. (**b**) Transversal complex array positional compensation.

The transversal C-scan array is similarly compensated by separating out the first/odd and second/even columns into separate arrays. Impedance data corresponding to a distance of a full coil pitch is discarded from the start of the odd array. Conversely, the opposite operation is performed on the even array where impedance data corresponding to a distance of a full coil pitch is discarded from the end of the array. This process is graphically illustrated in Figure 4b. Once all data are discarded, the odd and even arrays are interleaved together to make one C-scan array that is on the same positional grid as the axial C-scan array.

Once all data were collected and positionally compensated, oversampling is undertaken in the vertical direction of the array. No oversampling is performed in the horizontal scan direction as this is controlled adequately by setting the rotational speed and acquisition rate of the robot as described in the previous paragraphs. The oversampling is performed via linear interpolation of the raw impedance data. It was found that this linear interpolation was fast to implement and produced negligible errors with a maximum error of 2.12% and an average of 0.55% across both the axial and transversal datasets at 250 kHz.

By performing data compensation in this manner, a common spatial grid is established for each dataset configuration, enabling like-for-like comparison and further advanced post-processing techniques such as mixing of datasets.

2.3. Software Infrastructure

Extensive software infrastructure to control the eddy current Ectane device, as well as receive and process the acquired impedance data in real-time was developed and is documented in Figure 5. Literature has previously well documented the robotic software infrastructure required [31,43] and as a result, the work presented herein will focus on the eddy current software development effort.

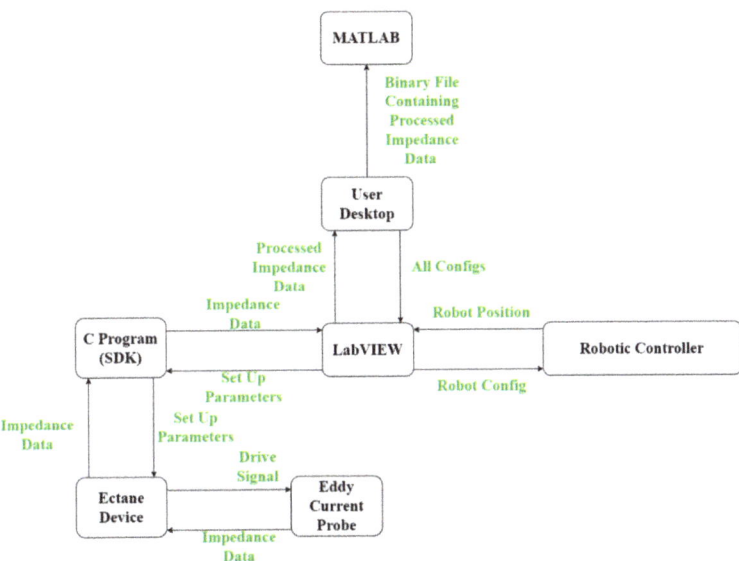

Figure 5. A flow chart showing the data transfer between different software and hardware elements.

In total 3 programs were developed: (1) A C program that houses the Eddyfi Ectane 2 SDK; (2) A LabVIEW program that receives, post-processes and plots impedance data in real-time as well as saving the data in a binary file format; and (3) A MATLAB reviewer program that reads in the binary file for further post-processing.

Both the C and LabVIEW programs are state machines. States within the C program are evaluated through a switch statement within the main while loop. In addition to the main while loop, the C program contains two threads that each have local host Transmission Control Protocol (TCP) connections. The first listens for standardised comma-separated string commands from LabVIEW and the other sends 32-bit impedance data from the Ectane device to the LabVIEW program. The same infrastructure with reverse logic is mimicked within the LabVIEW program through JKI state machines [44]. The standardised comma-separated string that is sent from LabVIEW is carried out in the following format:

```
state, IPAddress, configuration, acquisitionRate, gain,
freq1, voltage1, freq2, voltage2, freq3, voltage3,
freq4, voltage4, freq5, voltage5
```

As can be seen, there are 15 variables housed within the standardised string command. The first of which is the state that the C program should execute, and these are summarised below.

1. Do Nothing;
2. Connect to Device;
3. Set Up;
4. Balance;
5. Acquire Data;
6. Stop Data;
7. Disconnect from Device.

The second is the IP address of the Ectane device in order for the C program to connect to the Ectane device. Third is the configuration of the probe (i.e., will axial and/or transversal datasets be acquired? What probe is being used?). Next is the acquisition rate and gain of all Ectane channels. The final ten are the voltages and frequencies of each Ectane channel. As the Ectane device can acquire 5 datasets at different voltages and frequencies each of these must be specified even if some are unused.

The raw impedance data are received in the LabVIEW program as a series of 32-bit numbers and are immediately queued to be sequentially analysed in two additional threads. Using a 6 core, 2.6 GHz Intel i7-8850 H processor, it was found that the queueing of the received data was performed in 1 ms. As previously, these threads are implemented via two JKI state machines.

The first thread takes each 32-bit number and separates out the first and last 16-bits of data as these correspond to the imaginary and real impedance components. Additionally, the first thread reformats the impedance data into geometric order as the coils are pulsed in a pseudo-random fashion to prevent crosstalk caused by mutual inductance. Moreover, the first thread compensates for offset in coil excitation in the X-direction. Further details of this coil excitation compensation are provided in Section 2.2. It was found that this process was executed in 1 ms on a 6 core, 2.6 GHz Intel i7-8850H processor.

The second thread within the LabVIEW program is responsible for interpolation in the Y-direction between axial and transversal dataset configurations, oversampling, basic mixing of different datasets and live plotting of the impedance magnitude. As before, further details of this Y-direction interpolation and mixing of datasets are provided in Sections 2.2 and 2.4, respectively. Likewise, it was found that this process was executed in 5 ms on a 6 core, 2.6 GHz Intel i7-8850H processor. It is noted that the timings reported should be representative of any array used as the software infrastructure is built for the maximum number of elements, channel pairings, and number of frequencies.

This multi-threaded approach is illustrated in Figure 6 and provides data acquisition, positional compensation, and interpolation of impedance data whilst displaying various impedance magnitude C-scans in real-time to the user, all within the LabVIEW software environment with minimal 7 ms lag. The user can then select a directory to store the acquired data in a binary file format for future post-processing and analysis.

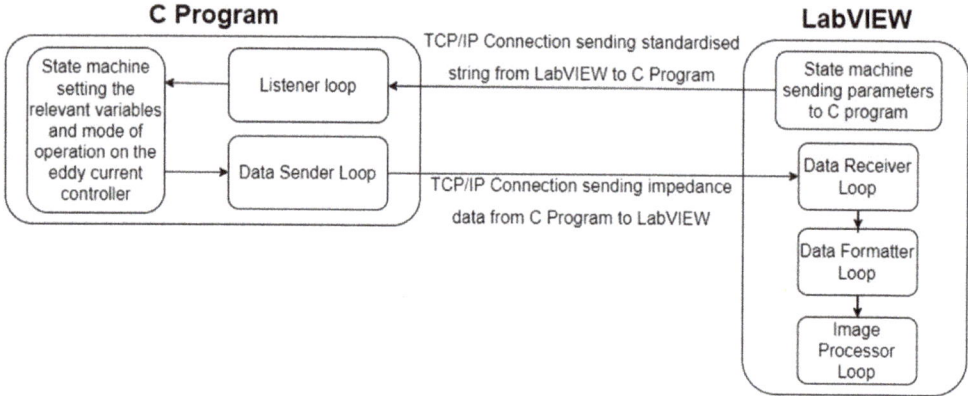

Figure 6. Illustration of the multi-threaded C and LabVIEW programs.

2.4. Image Enhancement of Impedance Data

It was shown in the literature that the impedance plane of the acquired data can be complex to interpret and variations in probe lift-off and wobble can commonly be mistaken as signals from defects [45,46]. Therefore, great care was taken in this work to minimise these adverse effects. Methods such as optimal probe design [47], multi-frequency excitation [45], and phase rotation [46] were shown to reduce such effects. Due to this work utilising commercial off-the-shelf (COTS) equipment, only multi-frequency excitation and phase rotation were performed. Multi-frequency excitation of 4 separate frequencies was conducted as the data were acquired and mixing of the datasets as described in Section 2.4.2 was performed in post-processing. Additionally, phase rotation was performed on the acquired C-scan datasets. All post-processing was performed via the MATLAB review application mentioned in Section 2.3.

2.4.1. Phase Rotation

The signature of adverse effects such as lift-off and wobble experience a phase difference in the response caused by a defect on the impedance plane. It is therefore common to phase rotate the data so that the response from the lift-off aligns with the horizontal axis of the display impedance plane, and plot C-scan images of the resulting vertical component of the impedance [48]. Due to the phase difference observed between the lift-off variations and that of a defect, the resulting C-scan will show any response from a defect clearly.

Mathematically, this is described in Equations (2) and (3). Equation (2) describes the resulting acquired impedance array from Section 2.3, and Equation (3) describes the mathematical operation performed to phase rotate the data by an angle, θ. This can be carried out at the point of acquisition or in post-processing. For this study, the decision was taken to phase rotate the data in post-processing to maintain maximum flexibility with the acquired data.

$$Z = R + iX \tag{2}$$

$$Z_{rot} = Z(\cos(\theta) + i\sin(\theta)) = (R + iX)(\cos(\theta) + i\sin(\theta)) \tag{3}$$

2.4.2. Mixing Eddy Current Datasets

As the impedance data were acquired onto a common spatial grid, mixing of datasets recorded under differing configurations or frequencies can be performed by superimposing the impedance C-scan data. This is graphically illustrated in Figure 7.

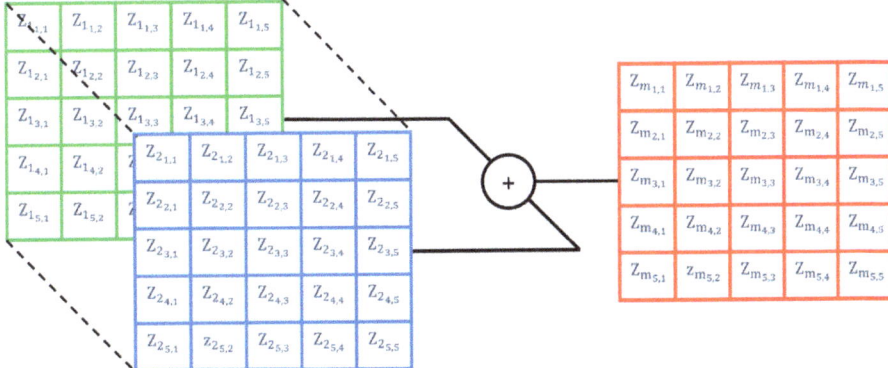

Figure 7. Illustration of mixing datasets Z_1 and Z_2 impedance data to make Z_m mixed data.

Two differing mixing methodologies were performed with the first being a simple sum and the second being a selective sum and average. As the name implies, the simple sum summates complex impedance datasets on a pixel-wise basis. For the selective sum and average, data above a defined noise floor were summated and everything below was averaged. The noise floor was defined as being 5 times the RMS values reported across a non-defective section of one of the impedance datasets to be mixed.

3. Results

Three 180-degree scans of the canister shown in Figure 2 were undertaken with both transversal and axial datasets being simultaneously acquired at frequencies of 250, 300, 400, and 450 kHz with an amplitude of 2 volts for each frequency channel, 30 dB of gain, an acquisition rate of 40 Hz, and a rotational speed of 1.72 deg/s. Each scan covered an area of 7687.1 mm² (array height of 32.625 mm × half the circumference of a 150 mm canister equating to 235.62 mm) making the final stitched image representative of an area of 23,061.3 mm². The interpolation was set to five, and the increments between half a coil pitch were specified at 20, giving a spatial resolution of 0.225 mm and 0.0563 mm in the vertical and horizontal directions, respectively. Positions were chosen for each scan so that they were acquired one array coil above each other with no overlap. The impedance data for all three scans were vertically stitched together and axial channel C-scans of the vertical impedance component from the impedance vector are shown in Figure 8. One of the stress corrosion cracks in the centre of the far-right column is highlighted. To the right of each C-scan, the impedance plane Lissajous for the highlighted defect is also shown along a horizontal cursor passing through the maximum intensity of the defect indication in the C-scan. It can be seen, that the impedance plane response of the same defect for different frequencies varies drastically in amplitude and phase due to the differing interaction depth of the eddy currents with the defect [46].

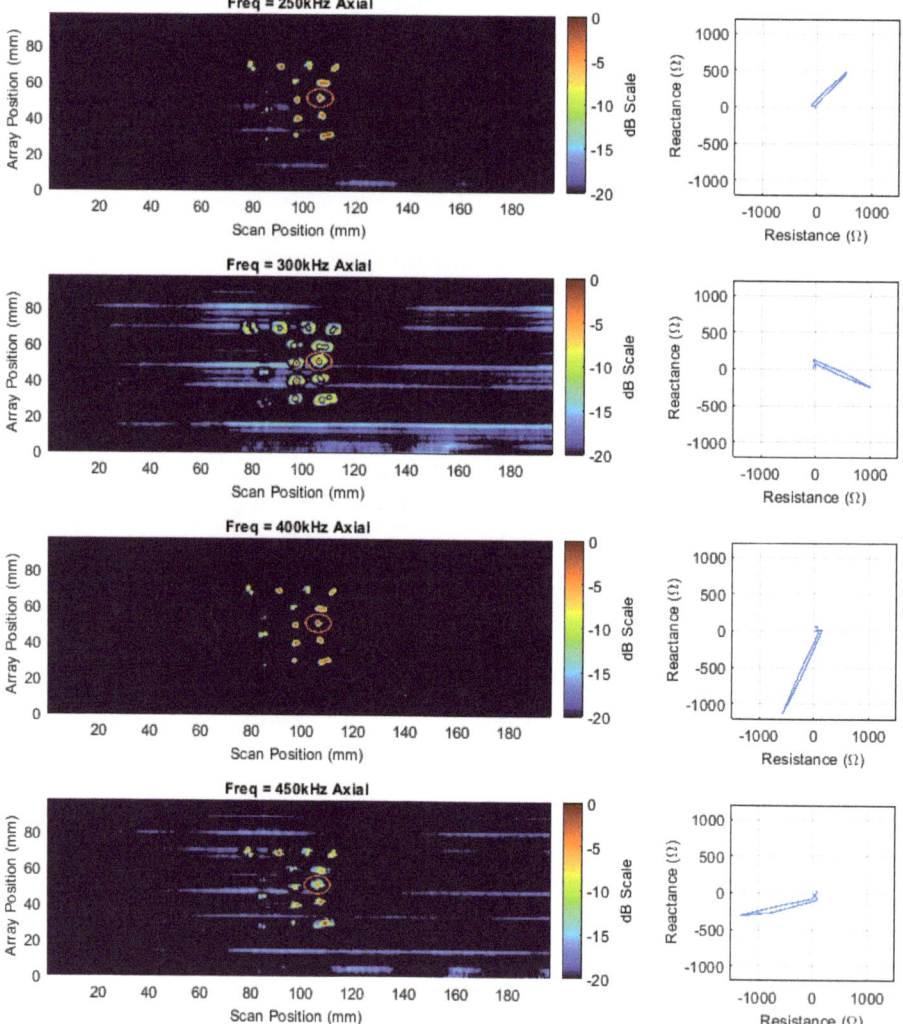

Figure 8. Axial vertical impedance component C–scan images at 250, 300, 400, and 450 kHz on a dB scale alongside impedance plane plots of the response from the highlighted defect.

Additionally, Figure 8 also shows that at 250 kHz and 450 kHz, the impedance plane contains a large horizontal component and as such the resulting image contains a large amount of noise. In order to compensate for this effect, the impedance data at each frequency were phase rotated so that the SNR of the highlighted defect was maximized – see Figure 9.

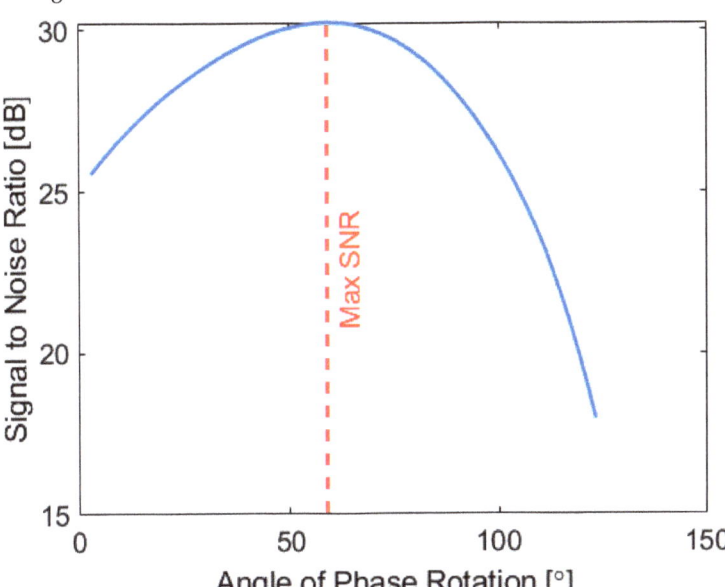

Figure 9. SNR vs. Angle of phase rotation for the axial dataset acquired at 250 kHz.

Figure 10 shows C-scan images of the optimised phase rotated axial data, while Table 2 denotes the SNR increases for both axial and transversal datasets at all frequencies recorded for the target defect. The increase in SNR for all defects is visually evident in Figure 10, and on average, the SNR was increased by 4.56 decibels for the targeted defect. This result illustrates the effectiveness that phase rotation can have on increasing the image performance of C-scans and the benefit of being able to flexibly perform such a task in post-processing.

Table 2. SNR Values of original and phase rotated data.

	250 kHz		300 kHz		400 kHz		450 kHz	
	Original SNR [dB]	Phase Rotated SNR [dB]	Original SNR [dB]	Phase Rotated SNR [dB]	Original SNR [dB]	Phase Rotated SNR [dB]	Original SNR [dB]	Phase Rotated SNR [dB]
Axial	25.02	30.23	20.19	31.22	29.86	31.27	21.11	31.27
Transversal	30.62	32.19	27.43	32.23	31.72	32.28	30.47	32.20

To further enhance image quality and reveal more about the nature of the defect, a mixing of different datasets, as described in Section 2.4.1, was performed. The optimised transversal and axial datasets at 250 and 450 kHz were mixed together, as the dissimilar frequencies would produce differing eddy current penetration depths and thus be influenced in differing manners. Equation (4) mathematically describes the penetration depth of an eddy current for a given material, where f is the frequency of the voltage being excited in the array coils in hertz (Hz), μ is the magnetic permeability of the component under test in henries per meter (H/m), and σ is the electrical conductivity of the component under test in siemens per meter (S/m).

$$\delta = \frac{1}{\sqrt{\pi f \mu \sigma}} \quad (4)$$

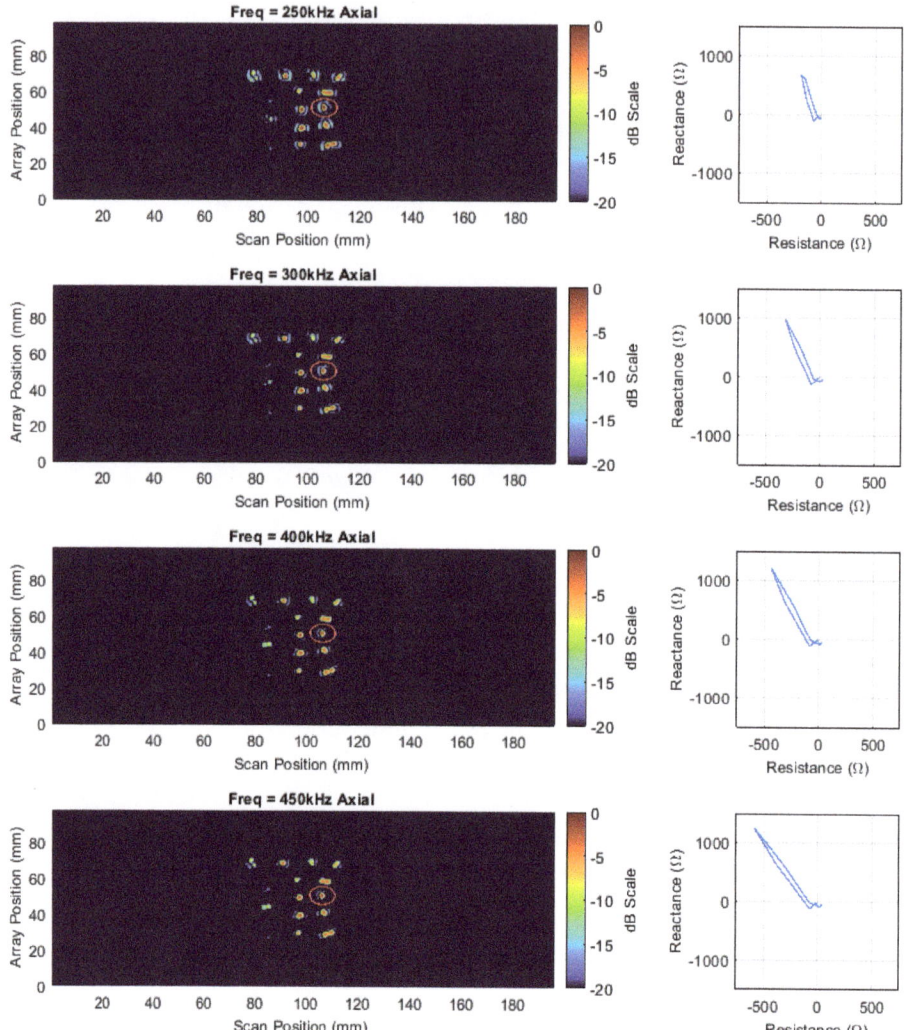

Figure 10. Phase-rotated axial vertical impedance component C−scan images at 250, 300, 400 and 450 kHz on a dB scale alongside impedance plane plots of the response from the highlighted defect.

For stainless steel, with an electrical conductivity of 1.08×10^6 S/m, and a relative magnetic permeability of 1.0025, a frequency of 250 kHz would produce a penetration depth of 0.967 mm, while a frequency of 450 kHz would produce a penetration depth of 0.721 mm.

The resulting mixed C-scan image is shown in Figure 11. Table 3 documents the SNR of the highlighted defect. As is shown in Table 3, the SNR of the defect for the simple sum approximates to be the average across all four datasets that contributed to the mixed image, and as such it can be said the imaging performance has not been improved by this mixing methodology. Interestingly, this is a result that is also observed in ultrasound when fusing multi-modal Total Focused Method (TFM) images [49]. By contrast, the selective sum and average technique were able to boost the SNR by an average of 1.19 dB, demonstrating an increase in imaging performance.

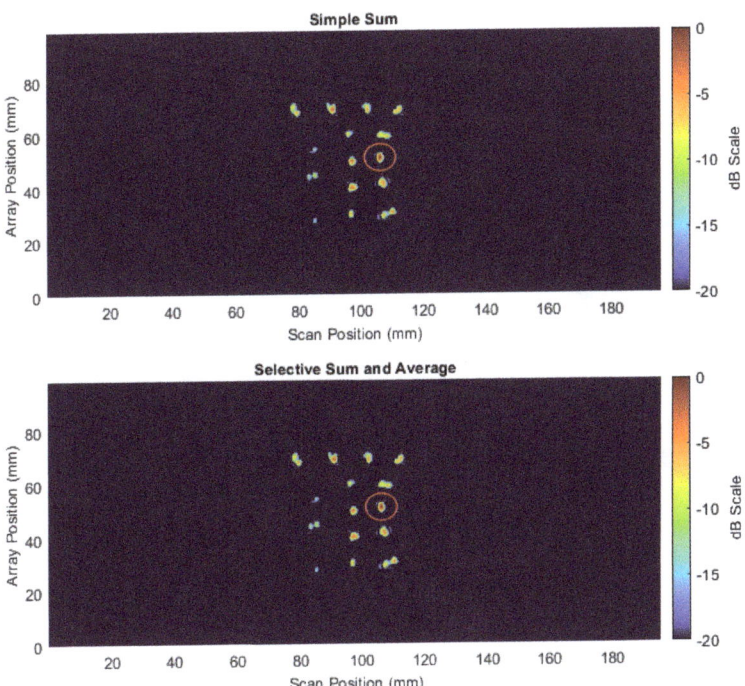

Figure 11. Mixed vertical impedance component C−scan.

Table 3. Mixed Image SNR.

	250 kHz	450 kHz	Mixed Data Simple Sum	Mixed Data Selective Sum
	Phase Rotated SNR [dB]	Phase Rotated SNR [dB]	Phase Rotated SNR [dB]	Phase Rotated SNR [dB]
Axial	30.22	31.27	31.85	32.66
Transversal	32.19	32.20		

It is acknowledged that in this study, SNR is the only metric being used to evaluate the eddy current detection system. A better metric would be a physical parameter related to the geometry of the defect itself (i.e., crack extent, crack depth) and whether this is better reflected in the mixing of datasets. As reported in the literature, this is a highly complex inversion problem, with successful inversions demonstrated on only simple geometries [50–53] or overall dimensions such as the depth or extent on complex defect geometries [54,55]. In all these studies, the defects were manufactured to specified geometries before eddy current testing which is somewhat removed from a real inspection scenario where prior knowledge of the defect geometry is not known. In addition, the sizing algorithms used vary drastically from defect to defect making the inversion of defect size somewhat deterministic and not well suited to automated deployment and analysis with which this paper is concerned. While the current system and signal processing cannot currently invert physical defect size, it was shown on another sample and different probe that is better suited to low-frequency operation, that the system is able to detect embedded defects ~3 mm below the inspection surface.

To understand more about the physical geometry of the highlighted stress corrosion crack, a macrograph was taken at 96 times zoom and is shown in Figure 12. It can be seen that the defect under inspection is a multifaceted stress corrosion crack. Due to its multifaceted nature, the interaction with the induced eddy current will be highly complex and therefore inversion of the physical geometry would be highly challenging. It is expected that for a simple linear defect, such as a fatigue crack, mixing of datasets would lead to benefits in defect characterisation even if the SNR was adversely affected. This issue is subject to future work and will be investigated by the authors at a later date.

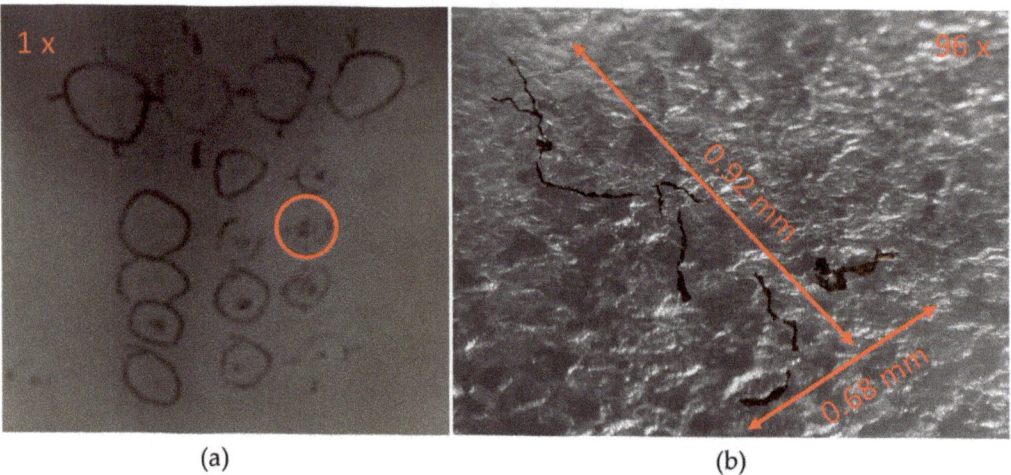

Figure 12. Photo of crack matrix and micrograph (**a**) Photo of crack matrix with the defect of interest highlighted in a red circle. (**b**) Micrograph of the defect of interest at 96× zoom with desaturated background.

4. Conclusions

This paper demonstrates for the first time how eddy current inspection with full image post-processing functions can be robotically deployed, showing a significant step closer to Industry 4.0 applications. Variations in the lift-off of the eddy current array were compensated for by the use of a PI control system and a force–torque sensor ensuring excellent low-noise coupling throughout the inspection. Extensive software infrastructure was developed that allowed for the eddy current data to be post-processed to enhance the generated images and reveal more about the nature of the defects under inspection.

The capability of the eddy current inspection system was demonstrated by inspecting a nuclear canister with a matrix of 16 stress corrosion cracks. Three 180-degree scans were conducted, gathering axial and transversal datasets at four different frequencies simultaneously—250, 300, 400, and 450 kHz—detecting 15/16 stress corrosion cracks. In the resulting data, one defect was highlighted, and various post-processing techniques were employed to increase the image quality. It was shown that, by phase rotation alone, the SNR could be increased by an average of 4.56 decibels. Dataset mixing was also attempted, and it was shown that a selective sum and average could boost the SNR by an average of 1.19 decibels. The multifaceted nature of the stress corrosion crack under inspection created a complex eddy current interaction, making it difficult to invert the physical geometry of the crack. It is expected that for simpler defect geometries a benefit in defect characterisation would be observed through dataset mixing.

This work demonstrated the detection of defects in real-time via eddy current data and showed the ability to further post-process the acquired data to enhance image quality. The benefit of being able to post-process the acquired data in such a manner should not be understated, and it is hoped that similar studies such as this can be used to further develop the post-processing of eddy current data to the standard achieved in ultrasonic NDT.

In future work, the authors plan to improve and progress this study by performing eddy current characterisation on multi-angled known defects and comparing the results to simulated datasets; exploring the use of machine learning to automatically classify and characterise defects; and lastly, exploring the fusion of ultrasonic and eddy current datasets.

Author Contributions: Conceptualization, E.A.F., S.M., E.N., C.L., M.V., D.L. and E.M.; Data curation, E.A.F. and M.M.; Formal analysis, E.A.F., M.M. and S.M.; Funding acquisition, G.B., R.B., E.M., A.G., G.P. and C.N.M.; Investigation, E.A.F. and E.M.; Methodology, E.A.F. and C.L.; Project administration, E.A.F.; Resources, E.M.; Software, E.A.F., E.N., C.L., M.V. and D.L.; Supervision, G.B., R.B., E.M., A.G., G.P. and C.N.M.; Validation, E.A.F., C.L. and E.M.; Visualization, E.A.F. and E.M.; Writing – original draft, E.A.F., G.B., R.B., M.M., S.M., E.N., C.L., M.V., D.L., E.M., A.G., G.P. and C.N.M.; Writing – review & editing, E.A.F., G.B., R.B., M.M., S.M., E.N., C.L., M.V., D.L., E.M., A.G., G.P. and C.N.M.. All authors have read and agreed to the published version of the manuscript.

Funding: This work is supported by the Research Centre for Non-Destructive Evaluation (RCNDE) on behalf of Sellafield Ltd. and NNL Ltd. in the UK under EPSRC Grant No. EP/L015587/1.

Informed Consent Statement: Not applicable.

Acknowledgments: The authors would like to thank Dirk Engelberg and his team at the Corrosion and Protection Centre within the Research Centre for Radwaste and Decommissioning at the University of Manchester for inducing the stress corrosion cracks on the canister samples.

Conflicts of Interest: The authors declare no conflict of interest.

References

1. Non-Destructive Testing (NDT) Market Size. Growth Analysis 2029. n.d. Available online: https://www.fortunebusinessinsights.com/non-destructive-testing-ndt-market-103596 (accessed on 7 June 2022).
2. Schulenburg, L. NDT 4.0: Opportunity or Threat? *Mater. Eval.* **2020**, *78*, 852–860. [CrossRef]
3. Mineo, C.; MacLeod, C.; Morozov, M.; Pierce, S.G.; Lardner, T.; Summan, R.; Powell, J.; McCubbin, P.; McCubbin, C.; Munro, G.; et al. Fast ultrasonic phased array inspection of complex geometries delivered through robotic manipulators and high speed data acquisition instrumentation. In Proceedings of the 2016 IEEE International Ultrasonics Symposium (IUS), Tours, France, 18–21 September 2016; pp. 1–4. [CrossRef]
4. Mineo, C.; MacLeod, C.N.; Su, R.; Lines, D.; Davì, S.; Cowan, B.; Pierce, S.G.; Paton, S.; Munro, G.; McCubbin, C.; et al. Robotic geometric and volumetric inspection of high value and large scale aircraft wings. In Proceedings of the 2019 IEEE International Workshop on Metrology for AeroSpace, Turin, Italy, 19–21 June 2019.
5. Mineo, C.; Vasilev, M.; Cowan, B.; MacLeod, C.N.; Pierce, S.G.; Wong, C.; Yang, E.; Fuentes, R.; Cross, E.J. Enabling robotic adaptive behaviour capabilities for new industry 4.0 automated quality inspection paradigms. *Insight-Non-Destr. Test. Cond. Monit.* **2020**, *62*, 338–344. [CrossRef]
6. Tabatabaeipour, M.; Trushkevych, O.; Dobie, G.; Edwards, R.; MacLeod, C.N.; Pierce, G. Guided wave based-occupancy grid robotic mapping. In *European Workshop on Structural Health Monitoring*; Springer: Cham, Switzerland, 2021; pp. 267–275. [CrossRef]
7. Javadi, Y.; MacLeod, C.N.; Pierce, S.G.; Gachagan, A.; Lines, D.; Mineo, C.; Ding, J.; Williams, S.W.; Vasilev, M.; Mohseni, E.; et al. Ultrasonic phased array inspection of a Wire + Arc Additive Manufactured (WAAM) sample with intentionally embedded defects. *Addit. Manuf.* **2019**, *29*, 100806. [CrossRef]
8. Javadi, Y.; Mohseni, E.; MacLeod, C.N.; Lines, D.; Vasilev, M.; Mineo, C.; Foster, E.; Pierce, S.G.; Gachagan, A. Continuous monitoring of an intentionally-manufactured crack using an automated welding and in-process inspection system. *Mater. Des.* **2020**, *191*, 108655. [CrossRef]
9. Lines, D.; Javadi, Y.; Mohseni, E.; Vasilev, M.; MacLeod, C.N.; Mineo, C.; Vithanage, R.W.; Qiu, Z.; Zimermann, R.; Loukas, C.; et al. A flexible robotic cell for in-process inspection of multi-pass welds. *Insight-Non-Destr. Test. Cond. Monit.* **2020**, *62*, 526–532. [CrossRef]
10. Mohseni, E.; Wathavana Vithanage, R.K.; Qiu, Z.; MacLeod, C.N.; Javadi, Y.; Lines, D.; Zimermann, R.; Pierce, G.; Gachagan, A.; Ding, J.; et al. A high temperature phased array ultrasonic roller probe designed for dry-coupled in-process inspection of wire + arc additive manufacturing. In Proceedings of the 47th Annual Review of Progress in Quantitative Nondestructive Evaluation, Minneapolis, MN, USA, 25–26 August 2020.
11. Javadi, Y.; Mohseni, E.; MacLeod, C.N.; Lines, D.; Vasilev, M.; Mineo, C.; Pierce, S.G.; Gachagan, A. High-temperature in-process inspection followed by 96-h robotic inspection of intentionally manufactured hydrogen crack in multi-pass robotic welding. *Int. J. Press. Vessel. Pip.* **2021**, *189*, 104288. [CrossRef]
12. Vasilev, M.; MacLeod, C.; Galbraith, W.; Javadi, Y.; Foster, E.; Dobie, G.; Pierce, G.; Gachagan, A. Non-contact in-process ultrasonic screening of thin fusion welded joints. *J. Manuf. Process.* **2021**, *64*, 445–454. [CrossRef]
13. Zhang, D.; Watson, R.; MacLeod, C.; Dobie, G.; Galbraith, W.; Pierce, G. Implementation and evaluation of an autonomous airborne ultrasound inspection system. *Nondestruct. Test. Eval.* **2022**, *37*, 1–21. [CrossRef]

14. Watson, R.; Kamel, M.; Zhang, D.; Dobie, G.; MacLeod, C.; Pierce, S.G.; Nieto, J. Dry coupled ultrasonic non-destructive evaluation using an over-actuated unmanned aerial vehicle. *IEEE Trans. Autom. Sci. Eng.* **2021**, 1–16. [CrossRef]
15. Watson, R.J.; Pierce, S.G.; Kamel, M.; Zhang, D.; MacLeod, C.N.; Dobie, G.; Bolton, G.; Dawood, T.; Nieto, J. Deployment of contact-based ultrasonic thickness measurements using over-actuated UAVs. In *European Workshop on Structural Health Monitoring*; Springer: Cham, Switzerland, 2020; pp. 683–694. [CrossRef]
16. Zhang, D.; Watson, R.; Dobie, G.; MacLeod, C.N.; Pierce, G. Autonomous ultrasonic inspection using unmanned aerial vehicle. In Proceedings of the 2018 IEEE International Ultrasonics Symposium, Kobe, Japan, 22–25 October 2018. [CrossRef]
17. Prashar, K.; Weston, M.; Drinkwater, B. Comparison and optimisation of fast array-based ultrasound testing. *Insight-Non-Destr. Test. Cond. Monit.* **2021**, *63*, 209–218. [CrossRef]
18. Zimermann, R.; Mohseni, E.; Vithanage, R.K.W.; Lines, D.; Foster, E.; Macleod, C.N.; Pierce, S.G.; Marinelli, G.; Williams, S.; Ding, J. Increasing the speed of automated ultrasonic inspection of as-built additive manufacturing components by the adoption of virtual source aperture. *Mater. Des.* **2022**, *220*, 110822. [CrossRef]
19. Pyle, R.J.; Bevan, R.L.T.; Hughes, R.R.; Ali, A.A.S.; Wilcox, P.D. Domain Adapted Deep-Learning for Improved Ultrasonic Crack Characterization Using Limited Experimental Data. *IEEE Trans. Ultrason. Ferroelectr. Freq. Control* **2022**, *69*, 1485–1496. [CrossRef]
20. Pyle, R.J.; Bevan, R.L.T.; Hughes, R.R.; Rachev, R.K.; Ali, A.A.S.; Wilcox, P.D. Deep Learning for Ultrasonic Crack Characterization in NDE. *IEEE Trans. Ultrason. Ferroelectr. Freq. Control* **2021**, *68*, 1854–1865. [CrossRef] [PubMed]
21. Pyle, R.J.; Hughes, R.R.; Ali, A.A.S.; Wilcox, P.D. Uncertainty Quantification for Deep Learning in Ultrasonic Crack Characterization. *IEEE Trans. Ultrason. Ferroelectr. Freq. Control* **2022**, *69*, 2339–2351. [CrossRef]
22. Cawley, P. Non-destructive testing—current capabilities and future directions. *Proc. Inst. Mech. Eng. Part L J. Mater. Des. Appl.* **2001**, *215*, 213–223. [CrossRef]
23. Sadiku, M.N.O. *Elements of Electromagnetics*; Oxford University Press: Oxford, UK, 2001.
24. Wheeler, H.A. Formulas for the Skin Effect. *Proc. IRE* **1942**, *30*, 412–424. [CrossRef]
25. Mackenzie, L.D.; Pierce, S.G.; Hayward, G. Robotic inspection system for non-destructive evaluation (nde) of pipes. *AIP Conf. Proc.* **2009**, *1096*, 1687–1694. [CrossRef]
26. Summan, R.; Pierce, G.; Macleod, C.; Mineo, C.; Riise, J.; Morozov, M.; Dobie, G.; Bolton, G.; Raude, A.; Dalpé, C.; et al. Conformable eddy current array delivery. *AIP Conf. Proc.* **2016**, *1706*, 170003. [CrossRef]
27. Morozov, M.; Pierce, S.G.; MacLeod, C.N.; Mineo, C.; Summan, R. Off-line scan path planning for robotic NDT. *Measurement* **2018**, *122*, 284–290. [CrossRef]
28. Zhang, W.; Wang, C.; Xie, F.; Zhang, H. Defect imaging curved surface based on flexible eddy current array sensor. *Measurement* **2020**, *151*, 107280. [CrossRef]
29. Kuka Industrial Robots High Payload Catalogue. n.d. Available online: https://www.kuka.com/-/media/kuka-downloads/imported/9cb8e311bfd744b4b0eab25ca883f6d3/kuka_pb_hohe_tl_en.pdf (accessed on 16 June 2022).
30. KUKA AG. n.d. DKP Two-Axis Positioner. Available online: https://www.kuka.com/en-gb/products/robotics-systems/robot-periphery/positionierer/dkp (accessed on 16 June 2022).
31. Vasilev, M.; MacLeod, C.N.; Loukas, C.; Javadi, Y.; Vithanage, R.K.W.; Lines, D.; Mohseni, E.; Pierce, S.G.; Gachagan, A. Sensor-enabled multi-robot system for automated welding and in-process ultrasonic NDE. *Sensors* **2021**, *21*, 5077. [CrossRef] [PubMed]
32. Mohseni, E.; Habibzadeh Boukani, H.; Ramos França, D.; Viens, M. A Study of the Automated Eddy Current Detection of Cracks in Steel Plates. *J. Nondestruct. Eval.* **2019**, *39*, 6. [CrossRef] [PubMed]
33. CatalogEddyfi-Surface-ECA-Probes.pdf. n.d. Available online: https://eddyfi.com/doc/Downloadables/CatalogEddyfi-Surface-ECA-Probes.pdf (accessed on 16 June 2022).
34. EddyFi. Ectane 2 Multi-Technology Tubing and Surface Test Instrument. Eddyfi 2019. Available online: http://www.eddyfi.com/products/ectane-2-test-instrument/ (accessed on 26 June 2019).
35. Holmes, C.; Drinkwater, B.W.; Wilcox, P.D. Post-processing of the full matrix of ultrasonic transmit–receive array data for non-destructive evaluation. *NDT E Int.* **2005**, *38*, 701–711. [CrossRef]
36. *ndtatbristol/mfmc*; Company Ultrasonics and Non-Destructive Testing Group: Caledonia, MI, USA, 2020.
37. Kroening, M.; Sednev, D.; Chumak, D. Nondestructive testing at nuclear facilities as basis for the 3S synergy implementation. In Proceedings of the 2012 7th International Forum on Strategic Technology (IFOST), Tomsk, Russia, 18–21 September 2012; pp. 1–4. [CrossRef]
38. Hyatt, N.C. Plutonium management policy in the United Kingdom: The need for a dual track strategy. *Energy Policy* **2017**, *101*, 303–309. [CrossRef]
39. ATI Industrial Automation: F/T Sensor Gamma IP65. n.d. Available online: https://www.ati-ia.com/products/ft/ft_models.aspx?id=Gamma+IP65 (accessed on 14 June 2022).
40. KUKA AG. n.d. KUKA KR C4. Available online: https://www.kuka.com/en-gb/products/robotics-systems/robot-controllers/kr-c4 (accessed on 16 June 2022).
41. KUKA AG. n.d. KUKA.SystemSoftware. Available online: https://www.kuka.com/en-us/products/robotics-systems/software/system-software/kuka_systemsoftware (accessed on 14 June 2022).
42. Ye, C.; Zhang, N.; Peng, L.; Tao, Y. Flexible Array Probe With In-Plane Differential Multichannels for Inspection of Microdefects on Curved Surface. *IEEE Trans. Ind. Electron.* **2022**, *69*, 900–910. [CrossRef]

43. Vasilev, M. Sensor-Enabled Robotics for Ultrasonic NDE. Ph.D. Thesis, University of Strathclyde, Glasgow, UK, 2021.
44. *JKI State Machine*; JKISoftware: Lafayette, CA, USA, 2022.
45. AbdAlla, A.N.; Faraj, M.A.; Samsuri, F.; Rifai, D.; Ali, K.; Al-Douri, Y. Challenges in improving the performance of eddy current testing: Review. *Meas. Control.* **2019**, *52*, 46–64. [CrossRef]
46. Shull, P.J. *Nondestructive Evaluation: Theory, Techniques, and Applications*; Dekker, M., Ed.; CRC Press: Boca Raton, FL, USA, 2002.
47. Auld, B.A.; Moulder, J.C. Review of Advances in Quantitative Eddy Current Nondestructive Evaluation. 34. *J. Nondestruct. Eval.* **1999**, *18*, 3–36. [CrossRef]
48. Mook, G.; Michel, F.; Simonin, J. Electromagnetic Imaging Using Probe Arrays. *Stroj. Vestn. J. Mech. Eng.* **2011**, *2011*, 227–236. [CrossRef]
49. Bevan, R.L.T.; Budyn, N.; Zhang, J.; Croxford, A.J.; Kitazawa, S.; Wilcox, P.D. Data Fusion of Multiview Ultrasonic Imaging for Characterization of Large Defects. *IEEE Trans. Ultrason. Ferroelectr. Freq. Control* **2020**, *67*, 2387–2401. [CrossRef] [PubMed]
50. Aldrin, J.C.; Sabbagh, H.A.; Annis, C.; Shell, E.B.; Knopp, J.; Lindgren, E.A. Assessing inversion performance and uncertainty in eddy current crack characterization applications. *AIP Conf. Proc.* **2015**, *1650*, 1873–1883. [CrossRef]
51. Aldrin, J.C.; Sabbagh, H.A.; Sabbagh, E.; Murphy, R.K.; Keiser, M.; Forsyth, D.S.; Lindgren, E.A. Model-based inverse methods for bolt-hole eddy current (BHEC) inspections. *AIP Conf. Proc.* **2014**, *1581*, 1433–1440. [CrossRef]
52. Shell, E.B.; Aldrin, J.C.; Sabbagh, H.A.; Sabbagh, E.; Murphy, R.K.; Mazdiyasni, S.; Lindgren, E.A. Demonstration of model-based inversion of electromagnetic signals for crack characterization. *AIP Conf. Proc.* **2015**, *1650*, 484–493. [CrossRef]
53. Liu, Y.; Tian, G.; Gao, B.; Lu, X.; Li, H.; Chen, X.; Zhang, Y.; Xiong, L. Depth quantification of rolling contact fatigue crack using skewness of eddy current pulsed thermography in stationary and scanning modes. *NDT E Int.* **2022**, *128*, 102630. [CrossRef]
54. Wang, L.; Cai, W.; Chen, Z. Quantitative evaluation of stress corrosion cracking based on crack conductivity model and intelligent algorithm from eddy current testing signals. *Nondestruct. Test. Eval.* **2020**, *35*, 378–394. [CrossRef]
55. Yusa, N. Development of computational inversion techniques to size cracks from eddy current signals. *Nondestruct. Test. Eval.* **2009**, *24*, 39–52. [CrossRef]

Article

Optimization Design and Flexible Detection Method of a Surface Adaptation Wall-Climbing Robot with Multisensor Integration for Petrochemical Tanks

Minglu Zhang, Xuan Zhang, Manhong Li *, Jian Cao and Zhexuan Huang

School of Mechanical Engineering, Hebei University of Technology, Tianjin 300130, China; zhangml@hebut.edu.cn (M.Z.); 201811201016@stu.hebut.edu.cn (X.Z.); 201921202068@stu.hebut.edu.cn (J.C.); lnzx0508@163.com (Z.H.)
* Correspondence: 2015038@hebut.edu.cn; Tel.: +86-138-0209-7213

Received: 27 October 2020; Accepted: 18 November 2020; Published: 20 November 2020

Abstract: Recently, numerous wall-climbing robots have been developed for petrochemical tank maintenance. However, most of them are difficult to be widely applied due to common problems such as poor adsorption capacity, low facade adaptability, and low detection accuracy. In order to realize automatic precise detection, an innovative wall-climbing robot system was designed. Based on magnetic circuit optimization, a passive adaptive moving mechanism that can adapt to the walls of different curvatures was proposed. In order to improve detection accuracy and efficiency, a flexible detection mechanism combining with a hooke hinge that can realize passive vertical alignment was designed to meet the detection requirements. Through the analysis of mechanical models under different working conditions, a hierarchical control system was established to complete the wall thickness and film thickness detection. The results showed that the robot could move safely and stably on the facade, as well as complete automatic precise detection.

Keywords: wall-climbing robot; passive adaptive mechanism; magnetic circuit optimization; flexible detection method

---

1. Introduction

With the rapid development of industries, an increasing number of spherical and cylindrical tanks have been used to store industrial products in the petrochemical field. Different degrees of damage in storage tanks have gradually emerged due to the open environment and natural aging, and regular maintenance has been adopted to ensure the safety of operation. However, traditional maintenance methods require a large number of humans and resources that are inefficient, costly, and dangerous [1–5]. Thus, developing a reliable and flexible wall-climbing robot has become a hot spot in the field of tank maintenance, as such a robot can realize the high precision detection of different detection modules under high risk and in complex petrochemical tanks [6–10].

At present, the adsorptive, moving, and detection mechanisms of wall-climbing robots have been extensively studied. Some typical robot systems have been developed and applied in various fields. The adsorption mechanism is the primary condition to ensure robot movement on a facade. Wall-climbing robots have different adsorption mechanisms for different working surfaces and moving modes. Numerous studies have revealed the following five adsorption modes: permanent magnet, electromagnetic, negative pressure, molecular force, and mixed adsorption [11–17]. Navaprakash et al. [18] used the principle of negative pressure adsorption to design an adsorption mechanism and verified its safe and stable adsorption on non-magnetic facades through software simulation. Chen et al. [19] designed a wall-climbing robot that uses a rotational-flow suction unit to realize climbing rough walls and overstepping small obstacles. Demirjian et al. [20] designed a caterpillar wall-climbing robot based on bionic principles that uses binder materials and breaks with traditional adsorption concepts. Seriani et al. [21] used wall-climbing robots on both sides of a wall to adsorb each other so as to realize the safe adsorption and stable movement on a non-magnetic wall. Wang et al. [22] optimized the magnetic circuit through the finite element analysis method and designed a new type of permanent magnet wheel with the same magnetic pole array arrangement that considerably improved the adsorption efficiency of the magnet. Wen [23] proposed an adjustable variable magnetic adsorption mechanism to realize the stability detection of a robot on the outer walls of storage tanks. Eto et al. [24] innovatively designed a two degrees-of-freedom (DOF) rotating magnetic attachment mechanism that maintains the optimal adsorption state of the magnet through passive adjustment and realizes safe and stable adsorption on different walls. Xiao et al. [25] designed a new steady-state permanent magnet adsorption operation mechanism to accomplish stable adsorption on complex facades. Fan et al. [26] combined electromagnetic and internal force compensation principles to realize the fast, controllable adsorption and separation of wall-climbing robots.

Many research institutions have developed a large number of wall-climbing robots for industrial applications based on the above adsorption mechanisms by combining mobile mechanisms and detection methods. By integrating viscous materials and a wheel-legged moving mechanism, Amirpasha et al. [27] innovatively proposed a wheeled foot-climbing robot that can achieve large obstacle crossing and wall transition. Wang et al. [28] creatively designed a bipedal, three-DOF wall-climbing robot to realize the detection of wind fan blades. Huang et al. [29] designed a crawler robot for ship detection by integrating a caterpillar structure and the magnetic adsorption mechanism that could realize the large-area detection of complex walls. Zhang et al. [30] designed a wall-climbing de-rusting robot for ship welds based on the visual recognition method of three-line laser structural light. Zhang et al. [31] developed a crawler wall-climbing robot to remove coatings based on high pressure water jet technology. In addition, numerous wall-climbing robots have been developed for petrochemical maintenance and other fields [32–37]. Mizota et al. [38] proposed a control method for the compliant motion of a wall-climbing robot based on propelling wave theory to realize stable and flexible movements on a façade by wall-climbing robots. Wu et al. [39] innovatively proposed a coordinated control method based on task trajectory tracking to realize the compliant detection of robots. Zhang [40] used an intelligent perception system to compliantly control a robot and to realize autonomous adaptive full-range detection over complex terrain. Song et al. [41] proposed an intelligent discrete trajectory tracking control algorithm based on the improved Dual-Heuristic Dynamic Programming (DHP) algorithm to solve the circular trajectory movement of a robot on a vertical wall.

Numerous wall-climbing robots have been developed and applied for petrochemical maintenance. However, current research is generally in the bottleneck state due to the limitations of reliable adsorption, surface adaptability, and detection devices, and the following three problems should be urgently solved. (1) Permanent magnet adsorption mechanisms have low magnetic energy utilization and adsorption capacity due to the limited transfer mechanism analysis of the multimedium magnetic circuit. (2) Moving the existing wall-climbing robots smoothly on curved surfaces with changeable morphologies is difficult due to insufficient studies on the passive flexible adaptive moving mechanism. (3) Achieving the vertical alignment of a probe for different detection modules while sticking to the facade is difficult for existing detection mechanisms, thus affecting detection effects and accuracy

Here, a wall-climbing detection robot that can realize multimode non-destructive testing on different walls is proposed on the basis of the above-mentioned problems. A high performance permanent magnet wheel was designed on the basis of magnetic circuit optimization to solve the safety adsorption problem, and the rapid demagnetization structure of the wheel was designed to facilitate the robot's removal from the wall after detection. Different from the traditional wall climbing mechanism with rigid connection, the wheels in this paper were flexibly connected with the moving mechanism to form a pseudopodia robot that could adapt to curved surfaces and move flexibly on the surfaces of spherical and cylindrical storage tanks. In order to improve the detection accuracy and efficiency of existing testing equipment, a flexible adaptive detection mechanism with multi-DOFs is proposed to passively adapt to wall surfaces by integrating a hooke hinge mechanism. A dynamic model of the wall-climbing robot was established on the basis of different working conditions to solve the momentum distribution problem of wheels under different motion modes. Through different process controls, the robot can use ultrasonic and eddy current probes to detect the thicknesses of wall and paint film, respectively. Experiments were conducted on a 5-mm-thick cylindrical tank surface to test the structure and detection capability of the robot. The experiments showed that the robot can move flexibly and stably on different facades. Simultaneously, the robot can accurately detect the thicknesses of walls and paint films by carrying different detection probes that can replace manual work to a certain extent.

The remainder of this paper is organized as follows. Section 2 introduces the structure of the detection robot, which mainly includes the magnetic adsorption moving mechanism and the passive flexible detection mechanism. Section 3 establishes mechanical analysis models for different working conditions and motion modes to determine the minimum adsorption and driving forces of safe and stable motions. Section 4 introduces the hardware composition of the control system and flexible detection process control flow with multiple detection capabilities. Section 5 presents the experimental process and analysis results. Section 6 provides several conclusions drawn from this research.

2. Introduction to Detection Robot

A wall-climbing detection robot adapted to different curvature walls was developed while considering the varied morphology of petrochemical tanks. The robot mainly comprised an adaptive moving mechanism, magnetic adsorption wheels, and a flexible detection mechanism with multi-DOF. The working environment of the robot comprises facades with different curvatures. Thus, solving the problems of the safe adsorption and stable movement of the moving mechanism, as well as the flexible adaptation and accurate measurement of the detection mechanism, was necessary. Therefore, a wall-climbing robot with flexible detection was developed. This robot can steadily adsorb and complete different detection tasks on different facades to meet the requirements of petrochemical tank detection. A high performance magnetic wheel structure that can be quickly demagnetized is also proposed. This structure coordinates the design of the multi-DOF moving mechanism to passively adapt to different curvature walls to ensure safe and stable movement. A flexible detection mechanism was designed in accordance with the operational requirements of the different detection modules by integrating rope pulling and a hooke hinge mechanism to realize the self-adaptive vertical alignment of the probe to adapt to different detection techniques. The detection robot can realize the precise movement and action of different detection process flows through a state control strategy and finally complete the wall detection tasks. The specific structure of the robot is shown in Figure 1.

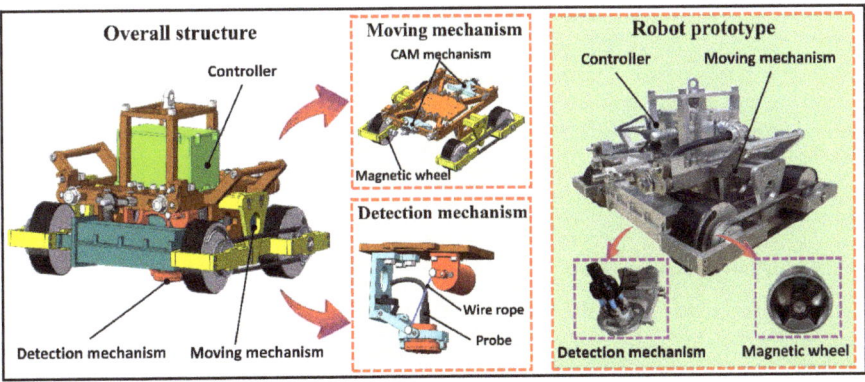

Figure 1. Overall structure of the wall-climbing detection robot.

2.1. Magnetic Wheel

2.1.1. Structural Design of the Magnetic Wheel

The permanent magnet adsorption mechanism is the crucial point in the design of a wall-climbing robot, because it is directly related to the safe absorption and stable movement on a wall surface. The robot movement is stable and the safety factor is large when the magnetic wheel adsorption capability is strong. However, the friction between magnetic wheels and the wall surface increases with the adsorption force and the resistance to be overcome in the movement is large, thus leading to a high driving torque. Simultaneously, detachment from the wall becomes difficult for the magnetic wheel after completing an avoidance detection task. Therefore, designing a lightweight wheel with strong adsorption was the pivotal technical problem to be solved in this paper. A new method using the combination of fan-shaped permanent magnet and yoke iron as the excitation source is proposed to improve the utilization rate of the magnet. At the same time, in order to detach the robot from the wall after completing the task, a fast demagnetization method was designed by using the lever principle. The specific structure is shown in Figure 2.

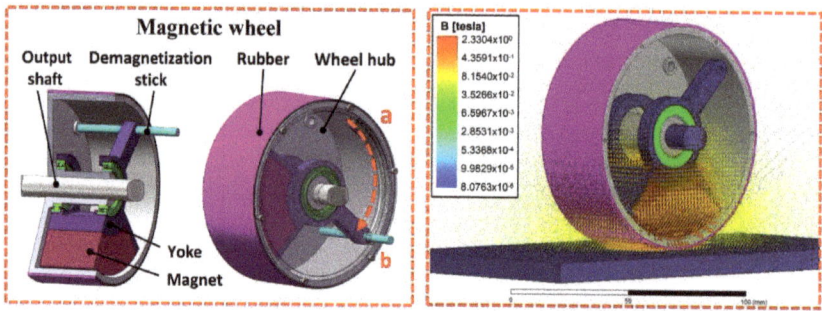

Figure 2. Magnetic wheel structure with fast demagnetization.

In the process of the adsorption force production of the magnetic wheel, most magnetic sensing lines come from a small part of the magnet close to the wall surface. Therefore, a radial magnetized fan magnet (Nd2Fe14B) was selected as the excitation source to reduce the weight and provide a strong adsorption force. Yoke iron was used to collect magnetic induction lines because of its high permeability that can reduce magnetic flux leakage and improve the utilization ratio of magnetic energy. Figure 2 shows that the fan magnet was placed in the suspension of the wheel, which could rotate relative to the wheel hub. When the output shaft transmits motion to the hub through a key, the permanent magnet always remains relatively still with the wall and does not rotate with the hub, which not only maintains a constant adsorption force but also avoids relative motion with the wheel. Actively reducing the adsorption force between the magnetic wheel and the wall, that is, the magnetic wheel demagnetization, is necessary to facilitate the robot detachment from the wall after the detection task. A small tangential force can be used in the adsorption state to force magnet rotation relative to each other, which can reduce the adsorption force between the magnet and the wall. Thus, a fast demagnetization mechanism was designed on the basis of the lever principle, which could facilitate magnet rotation around an output axis, thus completing the demagnetization.

2.1.2. Optimization of Magnetic Wheel

The magnetic wheel structure was optimized to obtain a high performance and lightweight magnetic wheel. The adsorption force of a magnetic wheel whose outside diameter and width are fixed is affected by the air gap h, the thickness of yoke iron H, and the shape of the magnet (the inner radius R_{in} and angle of the magnet θ). The electromagnetic field analysis software Ansoft was used to analyze the magnetic field strength of the permanent magnet and to determine the relationship between magnetic wheel parameters and the magnetic field strength to realize the lightweight of the magnetic wheel and ensure the reliability of adsorption. This analysis provided a reference for motion assessment and improved the magnetic utilization rate.

According to the principle of a single variable, the relationship between wheel adsorption force F and variable can be obtained by changing the air gap h, inner radius R_{in}, and angle of magnet θ. The simulation results are shown in Figure 3.

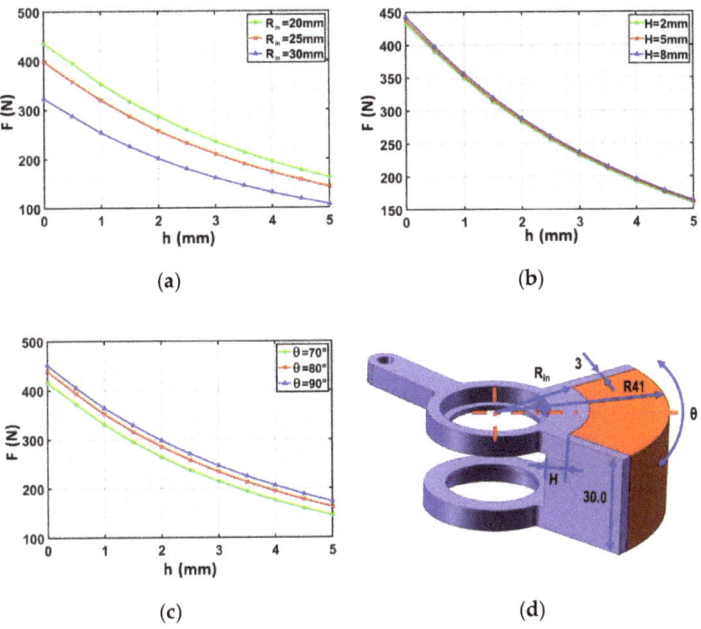

Figure 3. Influence of magnetic wheel parameters on the adsorption force: (**a**) The relationship between the adsorption force and the air gap height under different inner radius of the magnet, (**b**) the relationship between the adsorption force and the air gap height under different yoke iron thicknesses, (**c**) the relationship between the adsorption force and the air gap height under different magnet center angles, and (**d**) description of magnetic wheel structure and size.

The magnetic wheel adsorption force was found to be inversely proportional to the distance from the wall according to the information in the three above-mentioned figures; the adsorption force further away from the was found to be worse. Figure 3a shows that an increase in the inner radius R_{in} could lead to a contained high magnetic energy, a high magnetic field intensity that could be excited, and a strong adsorption force. Figure 3b indicates that the capability of the yoke iron to collect magnetic induction lines was found to increase with the yoke iron height H. This phenomenon complicates the magnetic saturation production and enhances the utilization ratio of magnetic energy products to improve magnetic field strength and adsorption force. Figure 3c reveals that the effective transfer area between the magnet and the wall surface was found to increase with the angle of the magnet θ, which improves the adsorption performance of the magnet. Considering the volume limitation of the wheel, the shape of the magnet was optimized in accordance with the functional relationship between the magnetic field strength and the geometric parameters of the magnetic wheel (h, R_{in}, θ, and H). Continuous nonlinear programming has the capability of using the response surface method to approximate the finite element response characteristics, which is very suitable for solving the optimization problem of finite variables. Therefore, the continuous nonlinear programming in Ansoft

Maxwell was adopted to optimize the three parameters of the magnetic wheel. The iterative process is complex and tedious but is commonly used; thus, comprehensively describing the solution process is unnecessary. Finally, a permanent magnet wheel with good performance was designed, and the specific mechanism size is shown in Table 1.

Table 1. Comparison of magnetic wheel characteristics before and after optimization.

Variate	Inner Radius R_{in} (mm)	Angle of Magnet θ (°)	Yoke Iron Height H (mm)
Before optimization	25	70	6
After optimization	20	80	5

The magnetic wheel adsorption experiment was conducted on an arc facade to test the adsorption capability of the magnetic wheel. The adsorption capacity of the wheel was tested in horizontal, vertical, and oblique states. The actual adsorption force was obtained by reading the maximum pull value of the magnetic wheel in the adsorption state on the wall through the dynamometer (the pull value of the wheel when it leaves the wall is the instantaneous maximum pull value). The influence of gravity in all cases was removed in the data recording process. The specific values are shown in Table 2.

Table 2. Adsorption force of the magnetic wheel under different conditions.

Times	Horizontal	Vertical	Oblique
First time	121 N	124 N	120 N
Second time	119 N	122 N	123 N
Third time	123 N	129 N	123 N

The actual adsorption force of the magnetic wheel could be obtained as 120 N by averaging the above values. The adsorption capacity can meet the requirement of safety adsorption of wall-climbing robot.

Thus far, a wheeled adsorption mechanism was innovatively designed. A lightweight permanent magnet wheel with strong adsorption capability was obtained through magnetic circuit optimization design and multivariable simulation optimization. The actual adsorption capacity of the magnetic wheel under different working conditions was then measured by experiments, and the adsorption performance of the magnetic wheel was verified.

2.2. Passive Adaptive Moving Mechanism

A detection robot works on a circular or spherical facade, and the adaptability of the moving mechanism to the complex wall is directly related to the movement safety and stability. Achieving the adaptability of small curvature tanks is difficult for traditional moving mechanisms, which will easily lead to slipping and instability, thus affecting work efficiency and operation safety. Moreover, realizing the stable movement of a robot on a wall becomes a problem. The possible instability of the detection robot was analyzed to solve this problem, which mainly includes the following two points. (1) A single front wheel is forced to leave a wall surface when a detection robot encounters an obstacle. Another wheel on the same side with a similar connection also leaves the surface due to the rigidity of the robot. This phenomenon directly leads to a sharp decline in the adsorptive capacity of the robot on the wall surface, thus making the robot prone to instability. (2) A robot's movement on the curved surface leads to an incomplete fitting of the angle between wheels and the wall. Ensuring enough adsorption force is difficult, and the decrease in contact area easily causes instability. Different from the traditional wheeled moving mechanism, we combined multi-DOF deformation concept to design an innovative moving mechanism with the ability for surface passive adaptation. The close contact between wheels and the wall was realized by passively adapting the fuselage component, thus ensuring the safe operation of the moving mechanism. The specific mechanism is shown in Figure 4.

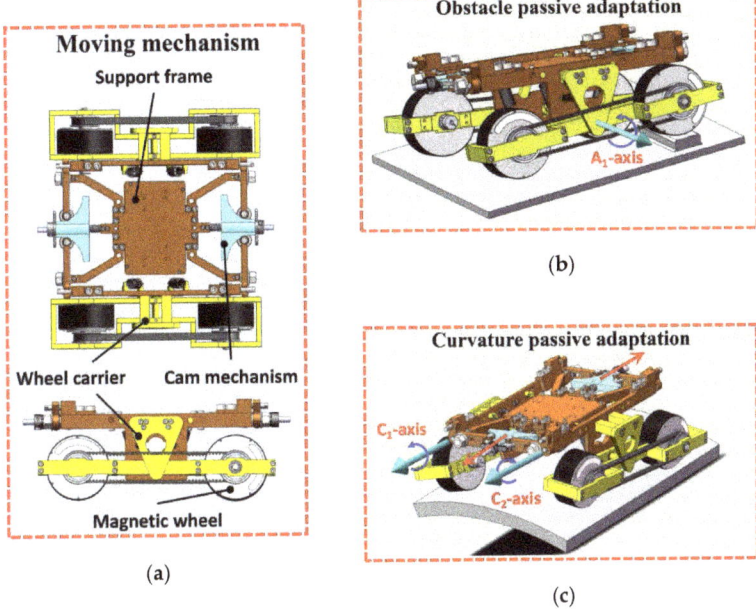

Figure 4. Pseudopodia flexible moving mechanism: (**a**) Moving mechanism structure, (**b**) obstacle crossing process, and (**c**) surface adaptation process.

A passive adaptive moving mechanism was designed in this paper to improve the adaptability of robots to facades and ensure their safe and stable movement. Figure 4 shows that the moving mechanism comprises the wheel frame, support frame, and cam mechanism. The wheels on the left and right sides are connected with the hand frame through the axes A_1 and A_2, respectively, and can rotate about the axes. The cam mechanism is fixed on the front and rear sides of the support frame by springs. The elastic deformation of the spring pushes the cam to move, which drives the wheel frame rotation around the axes C_1 and C_2; thus, both wheels fit vertically to the wall. Hence, the robot can be safely adsorbed on different curvature walls. The driving motors adopt diagonal arrangement and transfer power by using a synchronous belt to ensure the driving torque and simplify the control. Figure 4b shows that the right wheel frame rotates around axis A_1 when the unilateral wheel of the robot encounters obstacles to ensure that each wheel can be reliably adsorbed on the wall surface. This phenomenon avoids the first instability situation. Figure 4c shows that the cam mechanism is passively adjusted to drive the wheel frames on both sides moving to rotate around the axes C_1 and C_2 when the robot operates on the circular arc wall. Therefore, the magnetic wheel can closely contact the wall surface, which ensures stable and safe movements. Through the design of the above structure, the wheels on both sides of the robot can be flexibly adjusted with multi-DOF to ensure that each wheel can contact closely to different curvature walls and meet safety adsorption requirements.

2.3. Detection Mechanism

Nondestructive testing has always been highly recommended in the detection methods of petrochemical storage tanks. Ultrasonic and eddy current sensors are needed during the maintenance of petrochemical storage tanks to complete the thickness measurements of the wall and paint film. Different detection tasks require different technological processes, and the relative position between the detection device and the wall surface directly affects the detection effect and accuracy. Therefore, keeping the probe vertically aligned and close to the wall surface is necessary, while active and accurate real-time control increases the difficulty of control. Different from a traditional rigid detection mechanism, an underactuated passive adaptive detection mechanism was designed by integrating a hooke hinge mechanism to meet the precise detection requirements of different walls. The vertical alignment of probes is realized by the passive adaptation of the hooke hinge mechanism, and the probe is pressed tightly to the wall surface by the spring to meet the detection requirements. The specific architecture is shown in Figure 5.

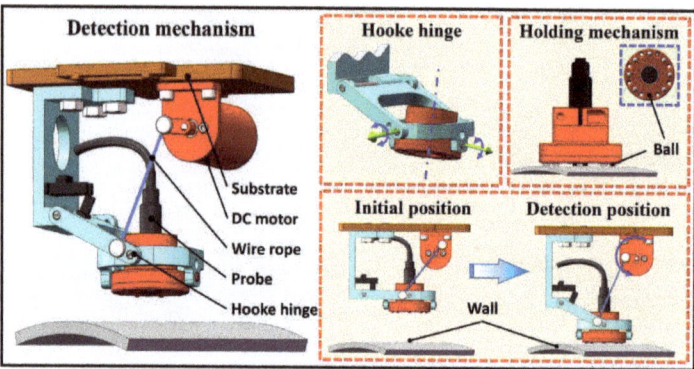

Figure 5. Flexible detection mechanism.

Figure 5 shows that the detection mechanism is fixed on the robot through the substrate. Hooke hinge structures enable a detection mechanism to have three DOFs, which can help detection mechanisms be perpendicular to kinds of complex wall surfaces. The torsion spring is installed on the rotary shaft of the hooke hinge mechanism. This hinge can provide torque force to press the probe on the wall surface to ensure the detection effect and improve detection accuracy. The hooke hinge mechanism is connected with the DC motor rocker arm through a wire rope. The DC motor rotates to lower the probe in the detection state. On the contrary, the DC motor rotates in reverse to lift the probe away from the wall in the non-detection state. The detection mechanism can be manually fixed into an L-shape by the locating pin after the removal of the wall-climbing robot from the wall surface. The holding mechanism in the hooke hinge mechanism is used to fix the detection probe, and different detection modules can be conveniently replaced to complete different detection tasks. In addition, eight stainless steel beads are installed uniformly on the probe holding mechanism to convert sliding friction into rolling friction to avoid damage to the detection wall. The above structure ensures close contact between the end of the detection mechanism and the surface. Therefore, detection efficiency can be guaranteed.

3. Mechanical Analysis

The weight and adsorption force of a robot directly affect the safety and movement flexibility when it runs on different facades. A robot must meet the safety requirements of different working conditions and movement modes in the process of continuous detection to realize the full domain detection of petrochemical storage tanks. Here, the critical failure states of the designed robot were analyzed through a mechanical model under different working conditions to obtain the minimum adsorption force of the magnetic wheel and ensure the safe and stable movement of the wall-climbing robot on a facade. Dynamic models were also established for different motion modes, and the robot and each wheel were analyzed to achieve the optimal momentum distribution and optimize the motion performance.

3.1. Statics Analysis

In the process of facade movement, a wall-climbing robot is prone to dangerous states, such as static sliding, vertical overturning, horizontal overturning, and oblique overturning. These states affect movement safety and flexibility. Thus, mechanics analysis on a robot must be conducted to determine the minimum adsorption force to ensure safe and stable movement. Here, a mechanical model was established for mechanics analysis, as shown in Figure 6.

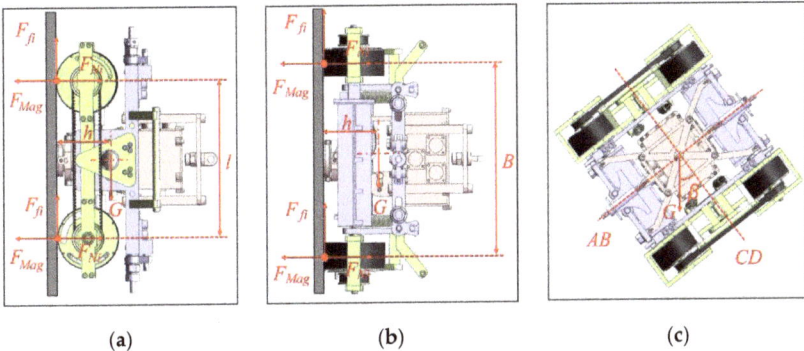

Figure 6. Static model of robot: (**a**) Vertical state, (**b**) horizontal state, and (**c**) oblique state.

Force and moment balance equations were established for the above states based on classical mechanics theory. In order to simplify the calculation process, we proposed the concept of safety factor to compensate for relatively small disadvantages such as cable weight and severe environment. The following static model of the robot was obtained.

$$\begin{cases} \sum_{i=1}^{4} F_{fi} = sG \\ \sum_{i=1}^{4} F_{Ni} = 4F_{Mag} \\ \sum_{i=1,3} (F_{Ni} - F_{Mag})l + sGh_c = 0 \\ \sum_{i=1}^{2} (F_{Ni} - F_{Mag})B + sGh_c = 0 \end{cases} \quad (1)$$

The meanings of the letters in the formula are shown in Table 3:

Table 3. Parameters in the mechanical model.

Symbol	Comment	Symbol	Comment
F_{fi}	Friction of the robot	G	Weight force of the robot
F_{Ni}	Support force of the wheel	s	Safety parameter
F_{Mag}	Adsorption of the wheel	μ	Static friction coefficient
l	Length of robot	B	Width of robot
ω	Angular velocity of turning state	h_c	Centroid height of the robot

The critical condition for the robot to be in a safe and stable state is that all magnetic wheels are on the wall surface, that is, constraints of support force and friction are present and the maximum static friction should be larger than the gravity component. Therefore, the value range of magnetic wheel adsorption force can be obtained as follows: $F_{Mag} \geq sG/4\mu$.

Figure 6c shows that when the robot is inclined to adsorb on the wall surface, the robot may flip around the AB or CD axes in this state. Gravity (G) can be decomposed into $G\sin\beta$ and $G\cos\beta$ along the direction of AB and CD. $G\sin\beta$ and $G\cos\beta$ were found to be less than G. Therefore, the calculated critical value of the safety adsorption force is less than the threshold of adsorption force when the robot is vertical and horizontal, as calculated above.

Therefore, the minimum adsorption force required by the robot was obtained to maintain static stability.

3.2. Dynamics Analysis

Dynamic analysis was conducted to obtain the optimal driving torque of each motor for the stable movement of the robot in different motion modes. The analysis of various motion modes revealed that the driving torque of other operation modes is less than or equal to that required for vertical upward straight or turning motion. Therefore, the dynamics analysis model of the robot was established in the two situations, and the best driving torque was obtained.

3.2.1. Dynamic Analysis in the Vertical Upward Movement

Dynamic analysis is similar to static analysis when running vertically upward. However, the existence of acceleration and the difference in the friction coefficient should be considered. Assuming that each wheel performs pure rolling motion without sliding, the mechanical model was obtained, as shown in Figure 7.

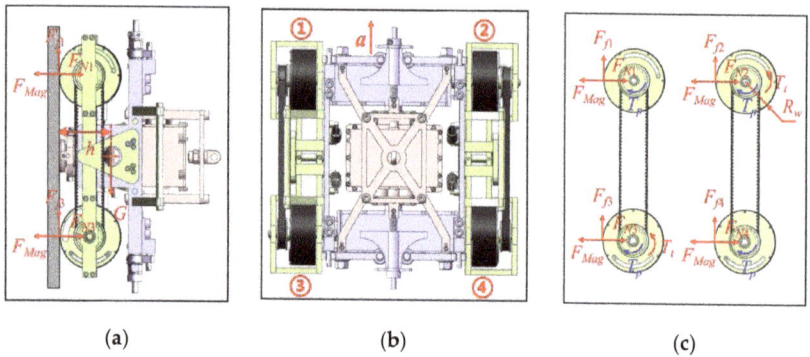

Figure 7. Dynamics model of the robot: (**a**) Side view of the vertical upward state, (**b**) main view of the vertical upward state, and (**c**) force analysis diagram of each wheel.

Rolling resistance was found to be generated due to the deformation of the rubber layer of the wheel, and its deformation was found to be small. Therefore, the moment of rolling resistance compared with other torsional moments could be ignored. The rolling resistance compared with other forces could also be disregarded due to the small rolling resistance coefficient. In order to enable the robot to overcome gravity and move stably, Formula (2) was established according to the principle of force balance, and then the required motor torque was solved.

$$2\frac{T_t}{R_w} - sG = \frac{sG}{g}a. \tag{2}$$

By simple derivation of Formulas (1) and (2), the required torque of the motor was calculated as follows:

$$T_t \geq (\frac{1}{2} + \frac{a}{2g})sGR_w. \tag{3}$$

The meaning of the letters in the formula is shown in the following Table 4:

Table 4. Parameters in the mechanical model.

Symbol	Comment	Symbol	Comment
V_s	Velocity of the center of mass	V_l	Velocity of the left two wheels
V_r	Velocity of the right two wheels	R_w	Radius of magnetic wheel
R	Radius of gyration	α	Angular acceleration of robot
T_P	Moment of resistance of wheel	T_t	Motor output torque

3.2.2. Dynamic Analysis in Steering

In this research, the wall-climbing robot as found to be able to achieve steering via different speeds of the wheels on each side. The angular speed and steering radius were, respectively, determined by the speed and direction of the wheels on both sides, as shown in Figure 8.

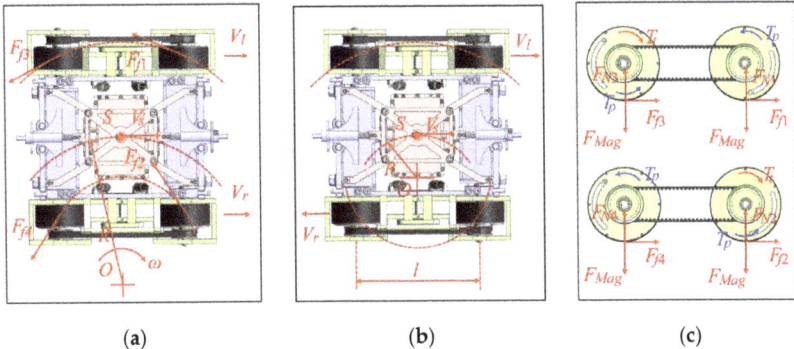

(a) (b) (c)

Figure 8. Dynamic analysis in steering: (a) Large radius turning state, (b) small radius turning state, and (c) dynamic analysis of each wheel.

Figure 8 shows the relationship between the rotation speed of wheels on both sides and the turning radius:

$$\begin{cases} \omega = \frac{V_l}{(R+B/2)} = \frac{V_r}{(R-B/2)} \\ \alpha = \dot{\omega} \end{cases} \quad (4)$$

The turning radius formula of the robot could be easily obtained according to the speed of the wheels on both side:

$$R = \frac{V_l + V_r}{V_l - V_r} \cdot \frac{B}{2}. \quad (5)$$

When $R > 0.5B$, the center of rotation is outside the robot (as shown in Figure 8a); when $R < 0.5B$, the center of rotation is inside the robot (as shown in Figure 8b). The condition of $R > 0.5B$ was taken as an example for force analysis. The horizontal to the vertical rotation of the wall-climbing robot was taken as the model to analyze the strained condition. The torque balance formula with point O as the center of rotation, as shown in Formula (6), was established to solve the required output torque of the motor. A mechanical model of the following turning states was obtained:

$$\frac{T_t}{R_w}(R+\frac{B}{2}) + \frac{T_t}{R_w}(R-\frac{B}{2}) - \sum_{i=1}^{4} F_{fi}\frac{l}{2} = J\alpha. \quad (6)$$

In combination with Formulas (4)–(6), Formula (7) could be obtained:

$$T_t \geq \frac{J\alpha + 2\mu F_{Mag} R_w l}{2R}. \quad (7)$$

The dynamics of the two motion modes of the robot, vertical upward motion and turning motion, were analyzed, and the equations were solved to find the motor torque range suitable for the stable operation of the robot.

4. Control System

This chapter introduces the hierarchical control system built by an industrial personal computer (IPC) as the upper computer, which uses IPC to realize the planning of the whole detection process and completes the detection task through the hierarchical control of each functional module. The control system mainly includes the precise motion control system of the flexible moving mechanism and the active adjustment control system of the flexible detection mechanism. Limited by the severe environment, wired control is used for remote controls to ensure the stability and accuracy of interactive information transmission. In addition to the basic function of flexible movement on the wall, the detection robot also needs to perform different detection modules for different detections. The robot can measure the thicknesses of the wall and paint film by, respectively, using ultrasonic and the eddy current probes. RS485 communication is adopted to complete the high-speed transmission of real-time detection information to realize multimodule and multimode coordinated detection, and the detection workflow is realized by using a distributed control system to monitor the close cooperation between the components. The specific control system structure is in Figure 9.

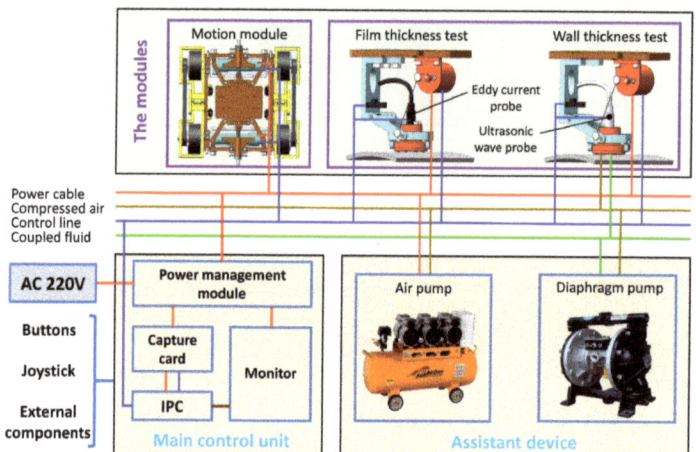

Figure 9. Hardware composition of the control system.

The robot is controlled remotely by external input devices, such as buttons in the control box. Control instructions are transmitted to the motors on both sides of the moving mechanism through the RS485. The motion pattern of the robot can be changed by adjusting the rotation speed and direction of the wheels on both sides. Simultaneously, inertial navigation information is used to adjust the running state of the motor to facilitate accurate movement. The distance between the detection device and the wall surface can be actively adjusted by controlling the motor steering in the flexible detection mechanism. The passive adaptability of the robot is used to ensure that the probe is perpendicular to the wall. Data obtained by probes are transmitted back to the main control unit in real-time via RS485 and displayed digitally. The robot can measure the thicknesses of the wall and paint film by, respectively, using ultrasonic and the eddy current probes. An air compressor and a diaphragm pump may be required during wall thickness measurement when using ultrasonic sensors to spray the coupling fluid near the probe to assist the robot in completing the wall thickness detection.

The control system of a wall-climbing robot is the key to motion and detection. The control can be divided into three parts: initialization, movement control, and detection control. The motion control realizes the flexible movement of the robot on the wall based on the feedback of the inertial sensor and the active remote control of the user. The detection control is allocated in accordance with different detection modules, which can be called for specific requirements. For example, measuring the position of marking points is necessary when the robot conducts the eddy current detection of paint film thickness while continuous monitoring and the coordination of coupling liquid when the robot conducts ultrasonic detection to wall thickness. The specific control flow chart is shown in Figure 10.

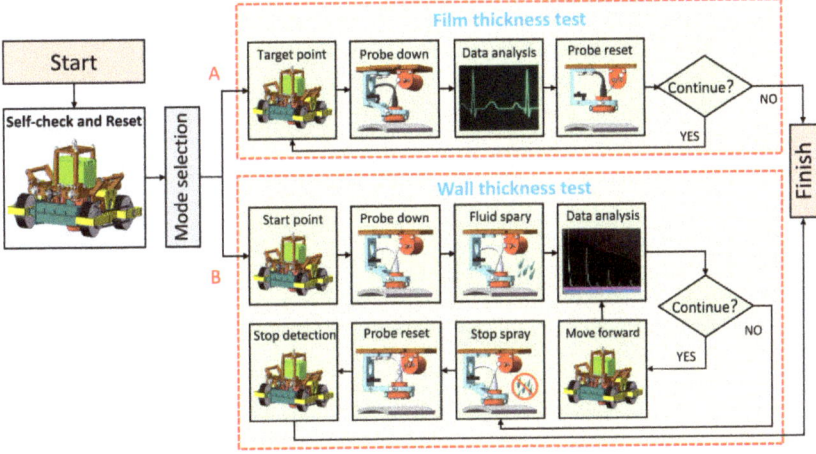

Figure 10. Robot detection workflow. (**A**) in the figure represents film thickness detection. (**B**) in the figure represents wall thickness detection.

The detection process can be analyzed as follows from the flow in Figure 10.

1. The robot is powered on to perform self-check and reset.
2. The user selects the detection mode:

    A: Film thickness detection.
    B: Wall thickness detection.

3. Different detection modes enter different detection processes:

    A: Film thickness detection:

    (1) The motors are driven to help the robot reach the initial detection point.
    (2) After the robot reaches the predetermined position, the DC motor in the flexible detection mechanism rotates forward to lower the detection probe. The probe cooperates with its passive adaptation mechanism to realize the vertical alignment of the probe.
    (3) After the sensor in the flexible detection mechanism confirms the detection position, the probe collects the paint film thickness information and transmits it back to the main control unit.
    (4) The DC motor reverses to lift the probe, and the robot completes the current position detection.
    (5) The main control unit checks the presence of a termination signal: if no termination signal is present, then the robot moves to the next detection point and repeat steps 2–5; if the termination signal is obtained, then the detection task is stopped.

    B: Wall thickness detection:

(1) Motors re driven to help the robot reach the initial detection point.
(2) After the robot reaches the predetermined position, the DC motor in the flexible detection mechanism rotates forward to lower the detection probe. The probe cooperates with its passive adaptation mechanism to realize the vertical alignment of the probe.
(3) The diaphragm pump sprays coupling fluid on the detection area to assist the detection task.
(4) The probe collects wall thickness information and returns it to the main control unit.
(5) The main control unit checks the presence of a termination signal: if no termination signal is present, then the robot continues to run and repeat steps 3–5; if a termination signal is obtained, then the diaphragm pump stops spraying coupling fluid and the DC motor reverses to lift the probe to stop the detection task.

4. The robot completes the detection task and resets.

5. Experiment

The authors of this paper designed a flexible and adaptive wall-climbing robot for film and wall thickness detection on curved wall surfaces by combining the magnetic wheel, flexible moving mechanism, and multi-DOF detection unit mentioned above. The key technical parameters of the robot are shown in Table 5.

Table 5. Keys technical parameters of the robot.

Items	Parameters
Weight	11 kg
Load capacity	9 kg
Maximum speed	10 m/min
Boundary dimension	400 × 400 × 300 mm
Communication mode	Wired (RS485)
Detection modes	Film/Wall thickness detection

This chapter discusses the movement and detection performance tests of the wall-climbing robot to verify the rationality and feasibility of the above-mentioned structure, control system, and the correctness of the mechanical theoretical analysis. The experimental facility mainly comprised the robot system and a cylindrical façade. Figure 11 shows a vertical circular steel plate, with a radius of 8 m and a wall thickness of 5 mm, which was used to simulate the tank environment. The detection robot system included the robot body, the control box, and auxiliary equipment. The air and diaphragm pumps of the auxiliary device provided coupling fluid for ultrasonic thickness detection. Operators controlled the movement and detection of the robot through the control cabinet to complete the detection task of the wall surface.

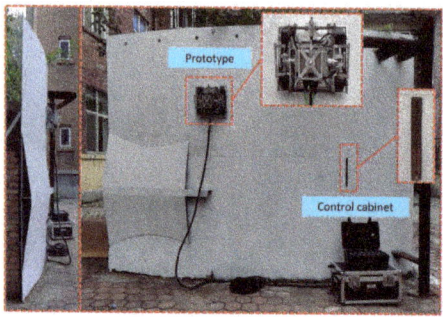

Figure 11. Experimental scene.

5.1. Movement Performance Test

Experiments on the vertical arc steel plate were conducted to test the stable adsorption and flexible precise movement capability of the robot. The performance of the robot was analyzed by monitoring the rotation speed of each wheel and the change in the position of the robot's center of mass during its vertical upward and horizontal circumferential movement on the arc plate. Among them, the wheel speed information was obtained by detecting the encoder information of the wheel motor, and the position information of the robot was obtained by the inertial navigation module mounted on the moving mechanism. The details are shown in Figure 12.

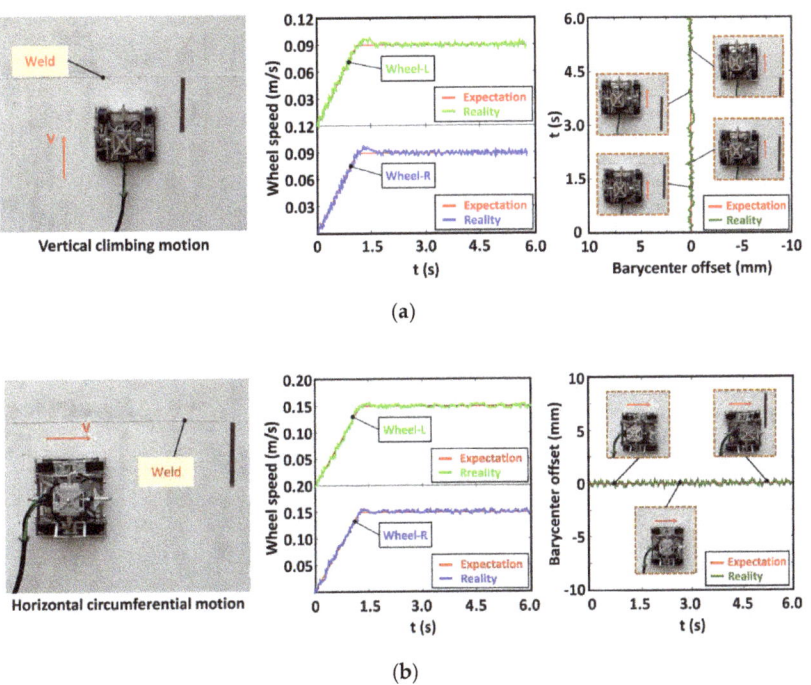

Figure 12. Experimental of wall motion: (**a**) Vertical upward climbing motion test and (**b**) horizontal circumferential motion test.

The robot could move safely and stably without slipping, falling, and other instabilities during the experiment. This finding indicated that the robot has good adsorption performance and adaptive capability. The synchronous belt is used to drive wheels on the same side. Therefore, the four-wheel robot could be simplified to the form of two wheels on right and left. The data in Figure 12a reveal that the rotation speed of the wheels on both sides during the vertical upward movement remained at 0.09 m/s despite slight fluctuations, and the position of the mass center did not deviate significantly in the horizontal direction. The data in Figure 12b show that the rotation speed of the wheels on both sides remained at 0.15 m/s in the horizontal circular motion of the robot, and the center of mass did not deviate significantly in the vertical direction. The above experimental data prove the steady and accurate robot movement on the circular steel plate.

A turn right movement experiment was conducted on the vertical circular arc wall to verify the movement flexibility. The robot was controlled to move from vertically upward to horizontally to the right, and the wheel rotation speed and the change of the mass center were detected in this process. The steel plate used in the experiment was expanded along the perimeter to intuitively understand the motion state, and the change curve of the mass center was drawn. The details are shown in Figure 13.

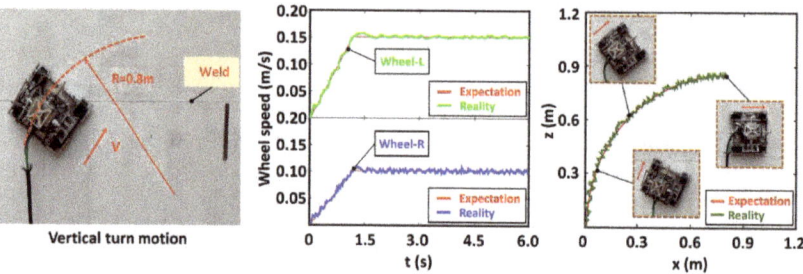

Figure 13. Turn motion test from vertical to horizontal.

The robot did not lose its stability and completed the right turning movement from the vertical direction to the horizontal direction in the above experiment. The left and right wheel rotation speeds were, respectively, set at 0.15 m/s and 0.09 m/s; therefore, the expected theoretical turning radius was 0.8 m. The figure above reveals that the rotation speed of the wheels on both sides slightly fluctuated up and down around the theoretical value. The adjusted trajectories show that the robot completed the turn, albeit with some deviation due to gravity.

5.2. Detection Accuracy Test

The wall-climbing detection robot could carry different detection equipment to detect a wall surface. Measurements of the paint film and wall thicknesses were taken as examples to conduct experiments to verify its detection capability. First, the test of film thickness was conducted. Mark points were set every 50 mm on the steel plate as the detection target points, and a handheld instrument was used to collect the detection information as the standard value. Then, the experiment data were automatically detected and recorded after setting the detection mode and the advance distance of the robot. The specific experimental process and two groups of test data are shown in Figure 14.

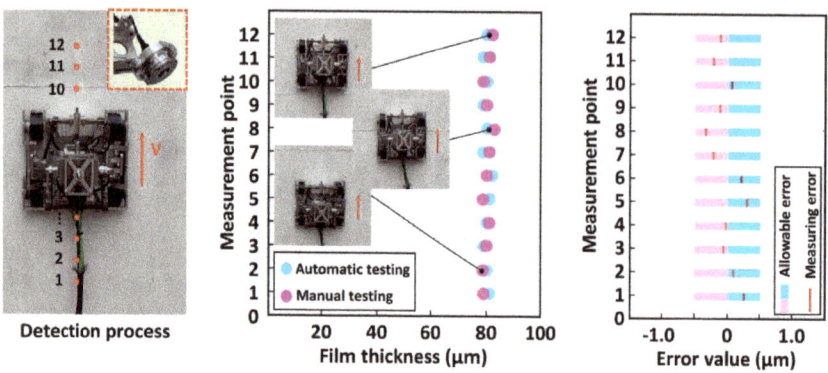

Figure 14. Measurement experiment of film thickness.

The robot could move accurately and complete the corresponding detection process during the experiment, thus finally achieving the measurement and data recording of paint film thickness. A comparison of the two above-mentioned sets of data revealed that the thickness of the paint film automatically detected by the robot was 80 um, which was close to that detected manually. Allowable error bands (±0.5 um) were set with industrial testing requirements after consulting relevant testing manuals, and all the measured values of the marking points were found to be within the allowable error band. The maximum error was 0.3 um, appearing at the eighth marker, which also met the detection requirements.

An auxiliary device was necessary for the thickness detection of steel plates to provide the coupling liquid for the detection robot. The thickness of the steel plate was measured at the continuously changing splicing steel plate, and experiments of manual measurement and automatic continuous measurement were also conducted. The specific experimental process and two groups of test data are shown in Figure 15.

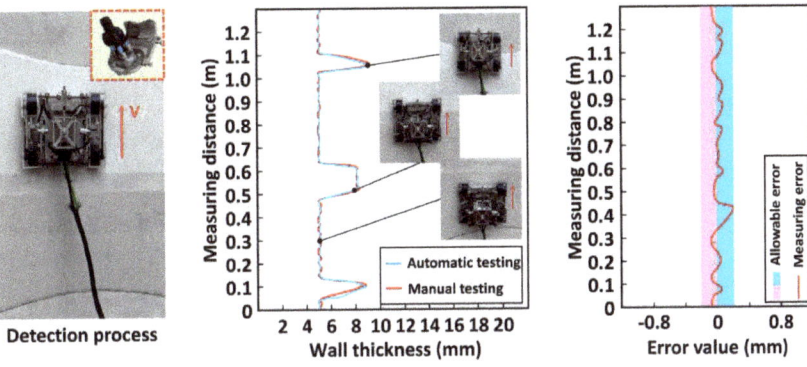

Figure 15. Measurement experiment of wall thickness.

During the experiment, the robot could effectively complete the detection process and conduct automatic detection continuously. Similarly, allowable error bands (±0.2 mm) that met the requirements of industrial testing were set for wall thickness measurement. Figure 15 intuitively shows that the robot could continuously detect the thickness of the steel plate, and its thickness changed from 5 to 8 to 5 mm, which was similar to the manual measurement result and met the detection requirements. The two kinds of test data considerably fluctuated at the welding seam due to the influence of welding quality and position deviation of measurement points. Thus, the detection accuracy problem at the welding seam was temporarily disregarded. Stable data could be collected at other locations, and the test results met the test requirements. The maximum error of measurement was +0.2 mm, which was also within the required error range and met the detection requirements.

The above experiments revealed that the designed wall-climbing robot could adapt to the curved wall and move safely, smoothly, and flexibly on the wall. Vertical alignment detection could be realized by carrying ultrasonic and eddy current probes and by cooperating with the passive adaptation of the multi-DOF flexible detection mechanism, and the detection tasks of wall and paint film thicknesses could be effectively completed. The experimental results showed that the robot could complete the task of accurate wall stability detection and realize the automatic surface detection of petrochemical storage tanks in the degree of movement.

6. Discussion and Conclusions

A wall-climbing detection robot that can adapt to tanks with different radii of curvature was designed to address the increasing maintenance and testing requirements of petrochemical storage tanks. The robot realizes the non-destructive detection of the wall surface and its safe operation through human remote and automatic controls. Different from the traditional adsorption mechanism, the fan-shaped permanent magnet, which added a yoke to collect the magnetic induction line, is used as the excitation source in this robot. This adsorption mechanism reduces the weight of the magnetic wheel, improves the utilization rate of magnetic energy, and ensures reliable adsorption. In order to solve the problem that the existing adsorption devices are difficult to detach from a wall after completing detection, an innovative fast demagnetization mechanism was designed by using the lever principle. Considering that the traditional rigid moving mechanisms are difficult to adapt to different tank wall environments (different curvatures and various obstacles), a flexible adaptive moving mechanism with multi-DOFs was innovatively designed. The multi-DOFs flexible deformation of the moving mechanism can adapt to a wall surface, which ensures a close fit between magnetic wheels and the wall surface. A flexible detection mechanism that was designed on the basis of the hooke hinge mechanism can quickly change detection equipment to meet the technical requirements of film and wall thickness detections. Through the passive adaptation of a multi-DOFs hooke hinge mechanism, the detection probe can always be perpendicular to the center and close to the wall surface, thus meeting the requirements of accurate detection. Considering various working conditions, the minimum adsorption force and the optimal driving force range of straight line and turning motion were calculated by establishing the mechanical model, which ensured the flexible and stable movement of the robot on an arc facade. Finally, the precise coordination control of each component is performed by the wired control to complete the detection task while considering the limitation of the severe environment.

The wall-climbing detection robot was found to be able to move stably on a façade by conducting experiments on a facade with a thickness of 5 mm and a radius of 8 m, which verified the adsorption capacity of the magnetic wheel. The robot could complete large-radius and in-situ turning movements, which verified the wall surface adaptability of the robot's moving mechanism. Through multisensor information fusion and multicomponent cooperation, the robot could complete the detection tasks of wall and paint film thickness detections by ultrasonic and eddy current sensors, respectively. The detection results also confirmed these findings. The experiment proved that the robot can complete the automatic wall detection task for petrochemical storage tanks.

Author Contributions: M.Z. provided supervision, formal analysis, and resources to the project. X.Z. designed all the experiments and subsequently drafted the manuscript. M.L. conceived the original ideas. J.C. contributed to the construction of the experiment platform. Z.H. conducted all the experiments and provided human resources. All authors have read and agreed to the published version of the manuscript.

Funding: This research was supported by the National Natural Science Foundation of China (Grant Nos. 61803142, 61733001, and U1913211), the Natural Science Foundation of Hebei Province (Grant Nos. F2018202210 and E2018202338), Hebei Science and Technology Agency Science and Technology Innovation Strategy Funding Project (Grant No. 20180603).

Conflicts of Interest: The authors declare no conflict of interest.

References

1. Cao, J.C. Risk Control of petrochemical storage Tanks. *Labor Prot.* **2015**, *8*, 104–106.
2. Zhou, J.; Liu, Z.X.; Dai, E.Q.; Zhang, G.M. Brief analysis on maintenance standards and management system of large storage tanks. *China Pet. Chem. Ind. Stand. Qual.* **2012**, *33*, 11.
3. Huang, H.W. Risk analysis and safety pre-evaluation of scaffold erection and removal. *China Sci. Technol. Prod. Saf.* **2017**, *13*, 36–41.
4. Zdravkov, L.; Pantusheva, M. Typical damage in steel storage tanks in operation. *Procedia Struct. Integr.* **2019**, *22*, 291–298. [CrossRef]
5. Li, H.Q.; Yang, Z.P.; Zhang, X.H.; Zhang, X.Z.; Li, Z.Y. Cleaning and maintenance of naphtala tanks. *Petrochem. Corros. Prot.* **2013**, *30*, 11–23.
6. Sattar, T.; Corsar, M.; James, R.; Seghier, D. Robotics Transforming the Future. In Proceedings of the 21st International Conference on Climbing and Walking Robots and the Support Technologies for Mobile Machines, Panama City, Panama, 10–12 September 2018; pp. 222–229.
7. Gao, F.; Fan, J.; Zhang, L.; Jiang, J.; He, S. Magnetic crawler climbing detection robot basing on metal magnetic memory testing technology. *Robot. Auton. Syst.* **2020**, *125*, 103439. [CrossRef]
8. Chen, X.; Wu, Y.; Hao, H.; Shi, H.; Huang, H. Tracked Wall-Climbing Robot for Calibration of Large Vertical Metal Tanks. *Appl. Sci.* **2019**, *9*, 2671. [CrossRef]
9. Abdulkader, R.E.; Veerajagadheswar, P.; Htet Lin, N.; Kumaran, S.; Vishaal, S.R.; Mohan, R.E. Sparrow: A Magnetic Climbing Robot for Autonomous Thickness Measurement in Ship Hull Maintenance. *J. Mar. Sci. Eng.* **2020**, *8*, 469. [CrossRef]
10. Gao, F.; Lin, J.; Ge, Y.; Lu, S.; Zhang, Y. A Mechanism and Method of Leak Detection for Pressure Vessel: Whether, When, and How. *IEEE Trans. Instrum. Meas.* **2020**, *69*, 6004–6015. [CrossRef]

11. Yang, W.; Zhang, W. A Worm-Inspired Robot Flexibly Steering on Horizontal and Vertical Surfaces. *Appl. Sci.* **2019**, *9*, 2168. [CrossRef]
12. Powelson, M.W.; Demirjian, W.A.; Canfield, S.L. Integrating Dry Adhesives and Compliant Suspension for Track-Type Climbing Robots. In Proceedings of the ASME 2019 International Design Engineering Technical Conferences and Computers and Information in Engineering Conference, Anaheim, CA, USA, 18–21 August 2019.
13. Xie, C.; Wu, X.; Wang, X. A Three-row Opposed Gripping Mechanism with Bioinspired Spiny Toes for Wall-climbing Robots. *J. Bionic Eng.* **2019**, *16*, 994–1006. [CrossRef]
14. Shao, J.; Li, X.; Zong, C.; Guo, W.; Bai, Y.; Dai, F.; Gao, X. A Wall-Climbing Robot with Gecko Features. In Proceedings of the 2012 IEEE International Conference on Mechatronics and Automation, Chengdu, China, 5–8 August 2012; pp. 942–947.
15. Yoshida, Y.; Ma, S. A Wall-Climbing Robot without Any Active Suction Mechanisms. In Proceedings of the 2011 IEEE International Conference on Robotics and Biomimetics, Karon Beach, Phuket, Thailand, 7–11 December 2011; pp. 2014–2019.
16. Zhou, Q.; Li, X. Experimental investigation on climbing robot using rotation-flow adsorption unit. *Robot. Auton. Syst.* **2018**, *105*, 112–120. [CrossRef]
17. Koh, K.H.; Sreekumar, M.; Ponnambalam, S.; Ponnambalam, S.G. Hybrid electrostatic and elastomer adhesion mechanism for wall climbing robot. *Mechatronics* **2016**, *35*, 122–135. [CrossRef]
18. Navaprakash, N.; Ramachandraiah, U.; Muthukumaran, G.; Rakesh, V.; Singh, A.P. Modeling and Experimental Analysis of Suction Pressure Generated by Active Suction Chamber Based Wall Climbing Robot with a Novel Bottom Restrictor. *Procedia Comput. Sci.* **2018**, *133*, 847–854. [CrossRef]
19. Chen, N.; Shi, K.; Li, X. Theoretical and Experimental Study and Design Method of Blade Height of a Rotational-Flow Suction Unit in a Wall-Climbing Robot. *J. Mech. Robot.* **2020**, *12*, 1–17. [CrossRef]
20. Demirjian, W.; Powelson, M.W.; Canfield, S.L. Design of Track-Type Climbing Robots Using Dry Adhesives and Compliant Suspension for Scalable Payloads. *J. Mech. Robot.* **2020**, *12*, 1–25. [CrossRef]
21. Seriani, S.; Scalera, L.; Caruso, M.; Gasparetto, A.; Gallina, P. Upside-Down Robots: Modeling and Experimental Validation of Magnetic-Adhesion Mobile Systems. *Robotics* **2019**, *8*, 41. [CrossRef]
22. Wang, J.D.; Xin, J.X.; Sun, A.Q.; Liang, M.X. Design and Optimization of permanent magnet wheel adsorption Device for wall climbing robot based on ANSYS. *Sci. Technol. Eng.* **2020**, *20*, 6931–6937.
23. Wen, J. Design and Characteristics of Tank Wall Crawling Robot. Ph.D. Thesis, Shanghai Jiao Tong University, Shang Hai, China, 2011.
24. Eto, H.; Asada, H.H. Development of a Wheeled Wall-Climbing Robot with a Shape-Adaptive Magnetic Adhesion Mechanism. In Proceedings of the 2020 IEEE International Conference on Robotics and Automation (ICRA), Paris, France, 31 May–31 August 2020; pp. 9329–9335.
25. Xiao, R.H.; Cheng, Y.X.; Jiang, Z.Z.; Zhang, R.J.; Wei, W. Design of robot adsorption mechanism for rust removal and wall-climbing on oil tank inner wall. *Mach. Des.* **2019**, *36*, 21–26.
26. Fan, J.; Xu, T.; Fang, Q.; Zhao, J.; Zhu, Y. A Novel Style Design of a Permanent-Magnetic Adsorption Mechanism for a Wall-Climbing Robot. *J. Mech. Robot.* **2020**, *12*, 1–30. [CrossRef]
27. Amirpasha, P.; Parviz, S.; Lu, J. A New Self-Loading Locomotion Mechanism for Wall Climbing Robots Employing Biomimetic Adhesives. *J. Bionic Eng.* **2013**, *10*, 12–18.
28. Wang, B.R.; Feng, W.B.; Luo, H.H.; Jin, Y.L.; Wu, S.Q. Design and stability analysis of bipedal three-degree-of-freedom wall-climbing robot on curved surface. *Robotics* **2014**, *36*, 349–354.
29. Huang, H.; Li, D.; Xue, Z.; Chen, X.; Liu, S.; Leng, J.; Wei, Y. Design and performance analysis of a tracked wall-climbing robot for ship inspection in shipbuilding. *Ocean. Eng.* **2017**, *131*, 224–230. [CrossRef]
30. Zhang, K.; Chen, Y.; Gui, H.; Li, D.; Li, Z. Identification of the deviation of seam tracking and weld cross type for the derusting of ship hulls using a wall-climbing robot based on three-line laser structural light. *J. Manuf. Process.* **2018**, *35*, 295–306. [CrossRef]
31. Zhang, F.; Sun, X.; Li, Z.; Mohsin, I.; Wei, Y.; He, K. Influence of Processing Parameters on Coating Removal for High Pressure Water Jet Technology Based on Wall-Climbing Robot. *Appl. Sci.* **2020**, *10*, 1862. [CrossRef]
32. Wang, Z.; Zhang, K.; Chen, Y.; Luo, Z.; Zheng, J. A real-time weld line detection for derusting wall-climbing robot using dual cameras. *J. Manuf. Process.* **2017**, *27*, 76–86. [CrossRef]
33. Vishaal, R.; Raghavan, P.; Rajesh, R.; Michael, S.; Elara, M.R. Design of Dual Purpose Cleaning Robot. *Procedia Comput. Sci.* **2018**, *133*, 518–525. [CrossRef]

34. Zhang, L.X.; Ji, W.G.; Li, T.X. Research on the Walking Control of the Pipeline Automatic Welding Trolley. *Weld. Technol.* **2016**, *44*, 69–70.
35. Chen, Y.; Mei, T.; Wang, X.J.; Li, F.; Liu, Y.W. Image detection and classification of bridge cracks based on wall-climbing robot. *J. Univ. Sci. Technol. China* **2016**, *46*, 788–796.
36. Cao, L.C.; Liu, X.G.; Jiang, X.M.; Zhang, H.; Wang, Z.M. Research on vertical welding technology of EH36 Marine high strength steel based on wall-climbing robot. *Precis. Form. Eng.* **2020**, *12*, 94–99.
37. Zhang, X.; Zhang, X.; Zhang, M.; Sun, L.; Li, M. Optimization Design and Flexible Detection Method of Wall-Climbing Robot System with Multiple Sensors Integration for Magnetic Particle Testing. *Sensors* **2020**, *20*, 4582. [CrossRef] [PubMed]
38. Mizota, Y.; Goto, Y.; Nakamura, T. Development of a Wall Climbing Robot Using the Mobile Mechanism of Continuous Traveling Waves Propagation Development of a Mechanism of Wave-Absorbing. In Proceedings of the 2013 IEEE International Conference on Robotics and Biomimetics (ROBIO), Shenzhen, China, 12–14 December 2013; pp. 1508–1513.
39. Wu, M.H. Research on Wheelfoot Non-Contact Magnetic Adsorption Wall-Climbing Robot for Welding Task. Ph.D. Thesis, Shanghai Jiao Tong University, Shanghai, China, 2014.
40. Zhang, L.; Ke, W.; Ye, Q.; Jiao, J. A novel laser vision sensor for weld line detection on wall-climbing robot. *Opt. Laser Technol.* **2014**, *60*, 69–79. [CrossRef]
41. Dian, S.; Fang, H.; Zhao, T.; Wu, Q.; Hu, Y.; Guo, R.; Li, S. Modeling and Trajectory Tracking Control for Magnetic Wheeled Mobile Robots Based on Improved Dual-Heuristic Dynamic Programming. *IEEE Trans. Ind. Inform.* **2020**, *99*, 1. [CrossRef]

Publisher's Note: MDPI stays neutral with regard to jurisdictional claims in published maps and institutional affiliations.

© 2020 by the authors. Licensee MDPI, Basel, Switzerland. This article is an open access article distributed under the terms and conditions of the Creative Commons Attribution (CC BY) license (http://creativecommons.org/licenses/by/4.0/).

Article

Soft-Tentacle Gripper for Pipe Crawling to Inspect Industrial Facilities Using UAVs

F. Javier Garcia Rubiales, Pablo Ramon Soria, Begoña C. Arrue * and Anibal Ollero

GRVC Robotics Laboratory, University of Seville, Avenida de los Descubrimientos, S/N, 41092 Seville, Spain; fragarrub@alum.us.es (F.J.G.R.); prs@us.es (P.R.S.); aollero@us.es (A.O.)
* Correspondence: barrue@us.es

Abstract: This paper presents a crawling mechanism using a soft-tentacle gripper integrated into an unmanned aerial vehicle for pipe inspection in industrial environments. The objective was to allow the aerial robot to perch and crawl along the pipe, minimizing the energy consumption, and allowing to perform contact inspection. This paper introduces the design of the soft limbs of the gripper and also the internal mechanism that allows movement along pipes. Several tests have been carried out to ensure the grasping capability on the pipe and the performance and reliability of the developed system. This paper shows the complete development of the system using additive manufacturing techniques and includes the results of experiments performed in realistic environments.

Keywords: UAVs; inspection; soft robotics

Citation: F. J. García-Rubiales, P. Ramon-Soria, B. C. Arrue and A. Ollero Soft-Tentacle Gripper for Pipe Crawling to Inspect Industrial Facilities Using UAVs. *Sensors* **2021**, *21*, 4142. https://doi.org/10.3390/s21124142

Academic Editors: Yashar Javadi and Carmelo Mineo

Received: 13 May 2021
Accepted: 10 June 2021
Published: 16 June 2021

Publisher's Note: MDPI stays neutral with regard to jurisdictional claims in published maps and institutional affiliations.

Copyright: © 2021 by the authors. Licensee MDPI, Basel, Switzerland. This article is an open access article distributed under the terms and conditions of the Creative Commons Attribution (CC BY) license (https://creativecommons.org/licenses/by/4.0/).

1. Introduction

The use of unmanned aerial vehicles (UAVs) has grown exponentially during the last decade. This growth has been associated with technological improvements, such as those in navigation systems and perception sensors.

Nowadays, there is an increasing interest in the use of UAVs for inspection and maintenance. At the moment, most inspection and maintenance tasks are carried out manually which exposes the operators to many dangerous situations. This paper focuses on facilities where there are tons of tubes and pipes that are required to be inspected, as can be found in the oil and gas sector.

In oil and gas production plants, some components degrade. The excessive corrosion of pipelines can lead to accidents, catastrophic failures, impact the environment, and affect plant availability. To prevent this situation, inspection processes such as wall thickness measurements are performed to ensure that plants have safe operating condition, or provide alerts for corrective actions if needed. These activities manually performed by operators. The main problem is that the structures to be inspected are in elevated locations at high temperatures or with toxic materials. This comes at a considerable cost to ensure the safety of inspection personnel and production outages.

By using UAVs, the operators are capable of inspecting inaccessible or dangerous zones without facing any risk. Furthermore, embedding sensors and cameras on the UAV allows them to perform more complex inspections. However, these operations using UAVs are still performed by manual control. The future of these applications relies on current research into the automation of these aerial systems.

For the accurate contact inspection of pipes with drones, landing gear is beneficial because it allows static contact to enable the UAV to perform measurements by coupling to pipes without causing any damage. Moreover, we are proposing a system that should also allow the robot to crawl along the pipeline. The soft gripper that is proposed in this paper is capable of having the necessary strength to hold onto the pipes, and move along them without causing damage.

Then, in addition to saving energy, compared to UAVs that can only fly, our hybrid (flying and crawling) locomotion system presented in this paper does not require the ability to accurately land on the inspection point because it can crawl after landing to be positioned where desired. The proposed system also has other benefits, compared to conventional crawlers, since its flying capability allows it to access places which would pose challenges for a human operator.

The idea is to use soft materials for the landing gear attached to the UAV as this is a safer alternative than other methods used in the state of the art. The use of soft materials in this area is a novelty when integrated into aerial robots. The problem is the difficulty of designing a lightweight gripper that is at the same time compact, energy-efficient and reliable.

Soft materials increase the adaptability of the holding system, while ensuring lower damage to the structures. This is an ideal solution for typical pipe inspection tasks in industrial facilities.

The ultimate goal is to have a system capable of crawling through pipes and inspecting them with ultrasonic sensors and make non-destructive testing (NDT) inspections. This kind of solution is very interesting for the industry and related service providers, as they can save costs, time and prevent undesirable accidents.

The rest of this paper is divided into five parts. The second section reviews the previous work. The third section describes the soft landing gear system, including the design, manufacturing process, and the operation flow of the landing gear and the soft limbs. The fourth section discusses the validation of the proposed system and the experiments carried out to validate its functioning. In the fifth section, the flight tests with the final setup is described and evaluated. Finally, our conclusions are drawn in the sixth section.

2. Related Work

Soft devices are currently being used in many areas of robotics because they provide advantages that the more traditional systems do not have, such as adaptability, compliance, better interaction with the environment and multi-functional end-effectors. Some examples of bio-inspired systems that can be used to interact with people are lightweight compliant arms with soft muscles that are pneumatically activated [1] and a pneumatic actuator can also be used to try to imitate the movements of a fish [2]. There are also soft grippers with muscles that are pneumatically activated to work in industrial environments interacting with humans [3].

There are many examples of materials and technologies, such as dielectric elastomer actuator (DEA), silicone-based elastomers, 3D-printed flexible actuators, or pneumatic actuators [4–7]. The main limitations associated with these soft-based actuators tend to be related to the complexity of the manufacturing process.

Other authors' examples are the high-contraction ratio pneumatic artificial muscle (HCRPAM) [8,9], prosthesis and grippers for manipulation [10], robots with elastomer actuators [11] and horticultural manipulation applications [12]. However, the main disadvantages of these actuators are the weight and space required by the pneumatic systems, which have motors and compressors with relatively high dimensions and weight.

Soft robotics are starting to be used in UAVs aiming to develop systems with capabilities to manipulate delicate objects, and to interact with people while flying. The use of soft materials is explored in [13] to become collision-resilient and increase its robustness. A special folding mechanism was investigated in [14]. DEA artificial muscle has also been used to try to simulate flapping wings, insects [15], or using a flexible membrane based on origami folding to preserve structural integrity during collisions [16].

The soft and compliant nature of the actuators ensures that soft robots are able to provide a safe interaction between the system and the facility to be inspected. Despite being lightweight structures, they are capable of achieving a high degree of freedom and a high force-to-weight ratio.

A large variety of robots have been designed to inspect pipes internally [17,18]. In fact, the most popular method of inspection is intelligent pigging [19,20], which makes use of devices equipped with sensors that navigate inside the pipes carried by the pipeline's fluid.

Robots have also been used to inspect pipes externally, usually called crawlers [21]. These robots commonly use magnetic wheels or tracks to move along the surface of pipes. This is very useful because the crawler can go underneath the pipe to make measurements, detecting possible leaks or corrosion. They are able to move freely over smooth surfaces, but in general, they cannot overcome obstacles and are limited to magnetic metal pipes.

The main locomotion alternative to these magnetic crawlers consists of an annular structure equipped with wheels, which obtain their adherence from a vacuum sucker.

UAVs can reach inaccessible areas faster than human operators or crawler robots. However, their flight endurance is very limited, and these robots are mainly used nowadays to perform visual inspections on structures by using different types of cameras (color, stereo, infrared), lasers, or other sensors. Authors in [22–24] proposed the use of UAVs to detect gas leaks and to monitor and map pipes. However, these solutions only allow for the visualization of surface damages.

Most recent research focused on the development of aerial robots that are able to not only perceive but also interact with the environment. This can be achieved by using robotic arms with several degrees of freedom (DOFs) that are attached to the UAV to interact and perform contact inspections [25,26]. These robots enable a new kind of application in which robots will be able to not only inspect but also perform the maintenance tasks at the industrial facilities [27]. These types of systems are called aerial robotic manipulators or aerial manipulators (AMs).

As related with previous perching mechanisms for UAVs, in [28], a single soft gripper was embedded at the bottom of a UAV to perch on pipes for inspection and maintenance tasks. Similar approaches have been taken in the rigid landing gripper of [29], the semi-soft perching system developed in [30] and the bio-inspired UAV with a soft landing gear that [31] used to land. However, these designs did not tackle the problem of moving along the pipe. The system presented in this paper allows crawling over the pipes to inspect them, saving time of flight.

3. Soft Landgear

In this section, the design of the landing gear and its characteristics are described. The section is split into two parts: the first one focuses on the description of forward-motion mechanism, while the second one focuses on the design and mechanical properties of the soft limbs.

The complete mechanism gear has been designed to allow the robot to crawl over the pipe. A compact and functional design that can be used for a variety of pipes because of the flexibility of the limbs is shown. Figure 1 shows the complete CAD design.

The gripper is manufactured with three different materials, which are: TPU, PLA and ecoflex. TPU (thermoplastic polyurethane) is a linear elastomeric polymer that can be used for 3D printing. Its greatest qualities are the flexibility and durability of the material. Polylactic acid (PLA) is a common plastic material in 3D printing. Finally, ecoflex is a cure silicone rubber compound. The advantages of ecoflex are that it mixes its components in equal parts to obtain a smooth and moldable silicone that can gain any shape and greatly increases the adhesion.

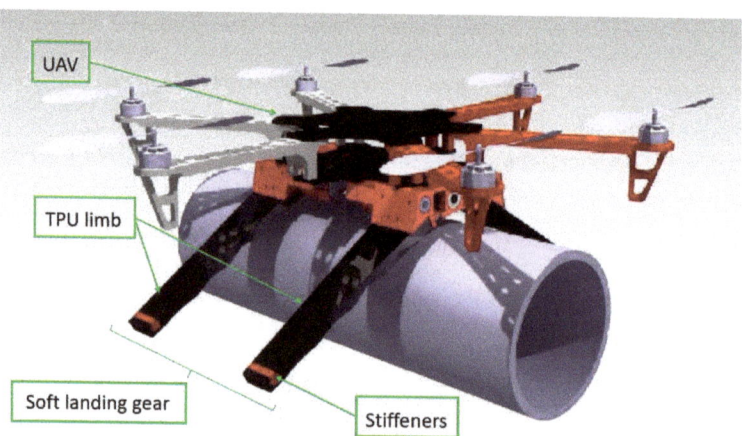

Figure 1. CAD design of the complete system.

3.1. Forward-Motion Mechanism

The forward-motion mechanism is a rigid core designed to be manufactured in PLA (as mentioned earlier). Additionally, all of the electronics are placed in this part because the rigid casing is more robust to possible impacts.

It has been designed to be flat on the upper side to make it easier to attach to different drones. This attachment is done with an embedded electromagnet. This special device is able to switch the magnetic field so that it can be enabled or disabled with a pulse. This can be used, for example, in situations where gas is detected or in emergency cases. This functionality is further explained in our previous work [32].

The mechanism is composed of two pieces: a fixed part that is attached to the drone and a mobile part that generates the forward movement. These two parts are symmetrical. The objective is to make a compact and robust design with the lowest possible weight so that it can be easily transported by the UAV, while the center of gravity does not significantly change when the UAV is moving.

Figure 2 shows the configuration of all the components. Three servomotors are responsible for all of the movements. Two of the servos are used to fold the soft limbs, which will be described later in Section 3.2. The other servo actuates as an endless screw that produces the forward movement. To restrict the torsion of the endless screw, two linear guides are located at the extremes of the rigid parts, which are attached with two 8 mm bars.

The lower part of the landing gear is a half-cylindrical section for the better adaptation to the pipes where it lands. The reference pipe size for this circular section is 160 mm, but the system can work in a diameter pipe range between 100 and 300 mm. On the sides, the system has two flaps protruding from the structure to attach the soft limbs.

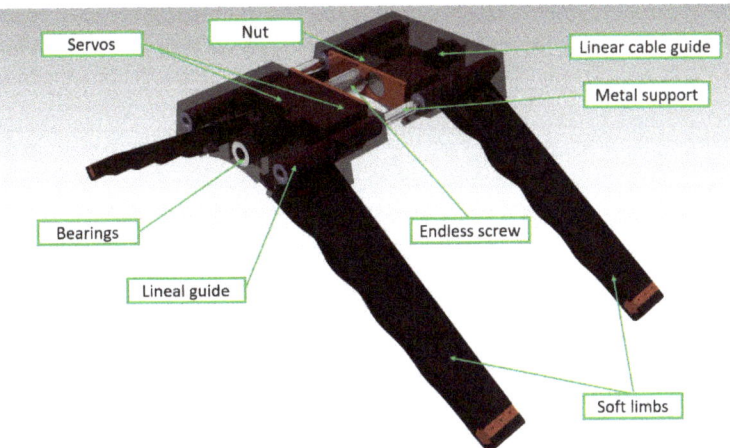

Figure 2. CAD view of the forward-motion system. The arrangement of the three servos, the worm gear, the linear bearings at the ends of the case and the guides to contracting the soft limbs can been seen.

3.2. Soft Limb Design

This section presents the design of the soft limbs. The points of interest of the soft-tentacle gripper are its capabilities of adapting to pipes of any diameter and absorbing the impact on the pipe while landing.

Each limb is made out of TPU. This rubber-like material provides the gripper with enough rigidity to retain its shape and maintain the exerting forces, but also enough elasticity to bend and adapt to different pipe shapes.

Another benefit of this material is that it can be used by a 3D printer, making it possible to easily iterate and develop different limb shapes. Finally, the tip of the limb is coated with silicone to increase the grip force of the system and obtain a softer contact with the pipe.

The main property of the gripper is that it is intrinsically compliant, allowing it to easily adapt to variations on the diameter of the pipe. Furthermore, weight is a crucial variable in aerial systems, where any additional payload means less flight time and it also affects the maneuverability of the UAV. Therefore, the landing system to be 3D printed has been designed to keep the weight to a minimum. Finally, the soft approach also offers more safety in the case of crashes.

The selection of the shape of the limbs has been decided after various tests and simulations, by observing the limb's deformation and finding the best adaptive shape for the pipe. In the studies carried out, the properties of the TPU material were analyzed. An example of simulation for the deformation of the limb is shown in Figure 3.

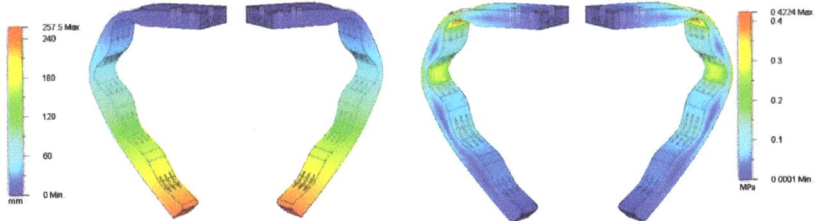

Figure 3. The left-hand image shows a study of the deformation for non-linear materials and the right-hand image shows a study of the stress.

To obtain a good grip and have the highest diameter range, the stiffness of the joint had to be taken into account. The stiffness formula for each joint is $j_i = EI/L$, where E and I are two constant parameters: E is the Young's modulus (the one used for the TPU is 100 N/mm^2), and I is the cross-sectional moment of inertia. L is the length between the nylon thread and the flexible segment. Figure 4 shows the final shape of the limb.

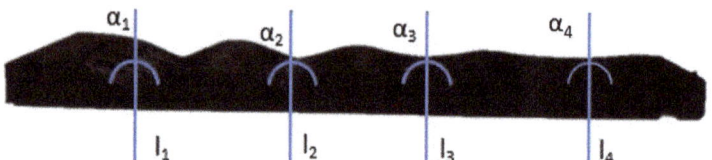

Figure 4. The image shows the final impression of the limb and the different angles chosen in its design.

The first tests were made with $j_1 = j_2 = j_3 = j_4$, where j_1 is the closest joint to the UAV and j_4 is the joint that is farthest away. This means that all lengths (L) were equal and, therefore, the same stiffness was generated at each joint of the soft limb. With this configuration, the limb started to bend first on the tip, which implied that it was not adjusting properly to the pipe, as shown in the left-hand example of Figure 5.

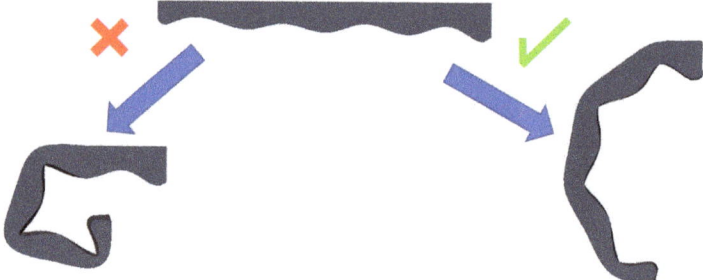

Figure 5. Example of the bad fold of soft limb and good fold when changing j_i parameters.

After this first attempt, the lengths L_i were changed, making them higher in the joints that are closer to the UAV, and reducing it progressively in the joints next to the tip; i.e., $j_1 < j_2 < j_3 < j_4$. With this approach, the tentacles bend better toward the shape of the pipe, increasing the grip of the system. The resulting stiffness generates the fold pattern that is shown in the right-hand example of Figure 5.

The following equation describes the recursive formula for the joint's stiffness [33]. The equations in (1) indicate the curvature generated by the limb joints:

$$j_n = \frac{1}{\theta_n}(F_t(m_n g(l_{cn} \cos \sum_{n=1}^{N} \theta_n - \frac{h}{2} \sin \sum_{n=1}^{N} \theta_n)$$
$$+ l_{an} \sin(\frac{\theta_n}{2}) + h))$$
$$R_{xn} = F_t - m_n g \sin \sum_{n=1}^{N} \theta_n \qquad (1)$$
$$R_{yn} = F_t \frac{\theta_n}{2} - m_n g \cos \sum_{n=1}^{N} \theta_n$$

These equations have been used to determine the joint deflections required to grip a circular cross-section of a particular radius R formed with $\theta i = \alpha$ angles. Figure 6 shows an example of how the limbs adapt to the circular section of a pipe, showing the different parameters used for the calculation. This design radius is considered to be optimal. Nevertheless, due to the softness of the framework, the gripper can envelope pipes of larger and smaller radius. Equations can be used to solve θi, giving as a result:

$$\begin{cases} \cos(\alpha) = \frac{R+H}{R+H+\delta} \\ \tan(\alpha) = \frac{a+L_{i-1}/2}{R+H} \\ \cos(\theta_i) = \frac{L_i}{2a} \\ \tan(\theta_i) = \frac{2\delta}{L_i} \end{cases} \quad (2)$$

$$\theta_i = \tan^{-1}\left(\frac{2(L_i^2 + L_{i-1} + L_i)(R+H)}{L_i(4(R+H)^2 - L_i^2)}\right) \quad (3)$$

By solving Equation (2), the system obtains δ and substitutes δ into $\tan(\theta_i)$ and Equation (3) is obtained.

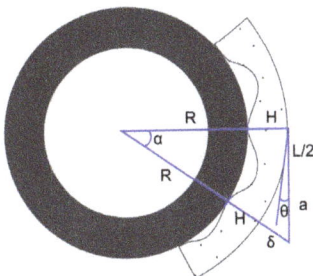

Figure 6. Model used to determine limb deflection in a pipe. H is the height of the limb, L is the length of the link, R is the radius of the pipe, θ is the deflection of the joint, and α is the angle between the contact points of two consecutive links.

Finally, by knowing each $\theta_i = \alpha$ angles for each joint, we can obtain β_i, the bending angle, from the following equations:

$$\begin{cases} S = \tan(\alpha)R \\ \beta_i + 2\psi_i = 180 \\ \cos(\psi_i) = \frac{S/2}{Z} \\ \sin(\beta_i/2) = \frac{S/2}{Z} \end{cases} \quad (4)$$

Figure 7 shows all the parameters used to solve the system of Equation (4):

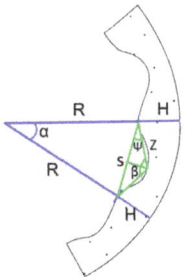

Figure 7. Model for the determinate used to determinate the β angle.

The conclusion was that to obtain the best fit possible of the limbs to the pipe, the shape of the limbs should suffer a progressive deformation from the base to the end of the limb. Table 1 shows the selected angles obtained for the selected radius R, the tendon tension F_t and the dimensions of the limbs.

Table 1. Final dimensions of the limb shape profiles.

Angle	β_1	β_2	β_3	β_4
degrees	132	136	168	175
length	l_1	l_2	l_3	l_4
millimeters	14	16	90	180

3.3. Complete Locomotion System

This subsection explains that all of the components conform to the locomotion system, and how they work. The locomotion system consists of three servomotors, which are embedded into the frame of the soft landing gear (as explained earlier).

Two of the servomotors are used to bend the soft limbs. The third is used to perform the linear displacement of the landing gear, which is made with an endless screw. Figure 8 shows the sequence of movements made by the soft land gear.

The first step to select the servomotors is to calculate the required minimum torque. The complete system weighs 3.25 kg, thus each limb will must exert ∼0.81 kg at its tip. Each motor has two limbs. For that reason, the final servo should exert at least 1.625 kg. By making this assumption, we grant that the gripper can hold the complete weight of the UAV in the worst scenario. Nevertheless, in most situations, this required strength will be lower, as part of the weight is held by the pipe, and the gripper only needs to prevent the UAV slipping laterally.

As shown in Figure 8, the motion runs as follows: two servomotors are hooked in pairs with the soft limbs using nylon threads, which allows them to open and close the limbs depending on the direction of the rotation of the motor. The third servomotor is responsible for the forward movement, using an endless screw that is connected to the motor at one end and to a nut at the opposite end. When the motor turns in one direction, the soft landing gear moves.

Figure 8. Example of movement sequence of the soft landing gear: (in **stage one**), it grips to the pipe; (in **stage two**), it opens the front limbs (blue case); (in **stage three**), it moves forward; (in **stage four**), it closes the front limb (blue case); (in **stage five**), it opens the rear limbs (black case); and finally, (in **stage six**), it moves the rear part.

3.4. Soft Limb Manufacturing Process and Assembly

This subsection will describe how we manufactured the limbs that are installed on the landing gear. Once the design has been carried out and has fulfilled the specifications, the manufacturing process of the limbs begins. One of the most important challenges is to be able to manufacture a soft part in a 3D printer, while making it easy to replicate and ensuring that the process is accurate.

All of the designs are made with TPU and a pair of PLA stiffeners, all produced on a 3D printer. These limbs are printed with different infill and printing patterns. The limbs were tested in the complete setup, until the one with more flexibility was selected. It was tested with a 10 percent infill which was very flexible and did not maintain the desired curvature when it was pulled by the nylon, then it was tested with a 20 percent infill which was very stiff and hardly allowed the limb to bend when it was in tension. The limb with an infill of 15 percent and a square printing patron was chosen as the optimal case.

Once the process of the impression of the TPU limb was finished, a PLA stiffener must be added to the tip to ensure that the tip does not deform. A stiffener is used to prevent losing strength at the tip when the nylon threads are contracted. After this, the different nylon strands that exert the force to deform and obtain the circular shape of the pipe, should be added to the limbs. These nylon threads are attached to the servomotor of the corresponding locomotion system and is then hooked at the end of the limb with the help of the PLA stiffeners at the end of the limb.

Finally, ecoflex is added. This elastomer is very flexible and rough. Ecoflex is applied to the tips of the limbs to improve adhesion to the pipe. To incorporate ecoflex with the TPU, molds were created using PLA with the shape of the limb tip and then were filled with ecoflex. The soft limb was then introduced into the molds, obtaining the silicone on the TPU. The ecoflex was cured for eight hours to obtain its physical qualities, and after this time, it was demolded. Figure 9 shows the final result of the operation.

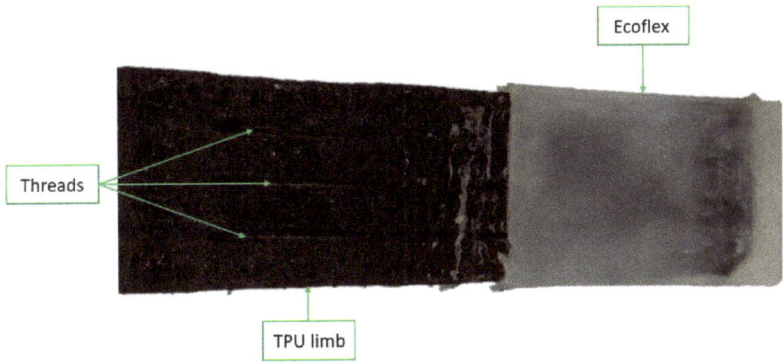

Figure 9. The final result when joining TPU with ecoflex. It can be seen that ecoflex is only applied to the tip to increase the adhesion in this area.

With all the parts manufactured, everything is assembled as follows. There are two parts. The first is the fixed part, where the two motors are located. One motor is responsible for closing and opening the limbs and the second motor operates the worm screw. This part has two holes at the top where the couplers are inserted. These are joined with metal bars. The second is the mobile part, which incorporates in its interior only a motor to drive the limbs. It also contains some superior holes where the linear bearings are inserted so that this part can be moved through the metallic bars. It also contains a nut in the frontal part where the screw without end will be connected.

The last step of the construction of the soft landing gear is to join these two parts with the worm screw and the metal bars. The metal bars are added to give consistency and to carry the weight of the landing gear so that the worm screw is not exposed to too much stress. It should be added that the motors in charge of opening and closing the limbs have a reel that is fixed to it and a bearing. The reel takes care of rolling up the nylon threads, and these threads pass along the limb and stay attached to the stiffeners.

4. Soft Land Gear Validation

This section presents the validation tests of the mechanical behavior of the soft landing gear. Experiments have been carried out to measure the force exerted by the limbs using different pipe sizes. The deformation when the limbs close over the pipe and the maximum slope ranges that the system can withstand without falling (laterally) are also studied.

It should be noted that the results in this section demonstrate the reliability of the design, as well as the actual capabilities of the gripper. These results can be extrapolated for manufacturing other customized landing gears for different pipe sizes and other payload requirements.

4.1. Pull Force

In this section, a comparison is made between various servos, checking the force that can be exerted both experimentally and theoretically. Then, one must choose the one that meets the design expectations and has the least weight and dimensions.

At first, the grip force is evaluated via experiments closing the claw on a test bench and measuring the force with a dynamometer. The dynamometer is hooked to the claws and pulled upwards to give real force exerted by the servomotor.

For those tests, the base of the limb was fixed and a force was applied at the tip. This force is equal to that exerted by the nylon on the limb.

The servomotor voltage that is given as an example of the datasheet is validated with experiments, comparing the force output for a dynamometer applying this voltage with the theoretical force obtained in the datasheet.

A comparison has been made with three servos, the Feetech FS5103B, the Feetech SCS15 and the Feetech FT6325M. According to the technical specifications, the first servo has a voltage operation ranging from 4.8 to 6, the second has a voltage operation ranging from 6 to 8.4 and the third also has a voltage operation ranging from 6 to 8.4, whilst the force range obtained for each servo is as follows: for the first, the force ranges from 0.5 to 0.7 kg; for the second, the force ranges from 2.2 to 2.9 kg; and for the third, the force ranges from 2.8 to 3.6 kg.

Figure 10 shows that the theoretical force is greater than the experimental force, which was expected, due to the normal mechanical losses. This loss is lower than the 4%, being bigger with lower voltages and lower with higher voltages. It is observed that the servo 1 graph does not perform the specifications while the second and third graphs do, and the third one more than complies with the specifications. Finally, the second one is chosen as it meets the design requirements and the servo has a lower weight.

After analyzing this information, we chose to use the Feetech SCS15 servomotors because they have enough force to move the landing gear and hold on to the pipes. In addition, they have a serial bus connection in which the three motors can be connected at the same time with the same bus and each motor can also be selected according to the motor ID. In Section 4.1, we will explain why this servo was selected.

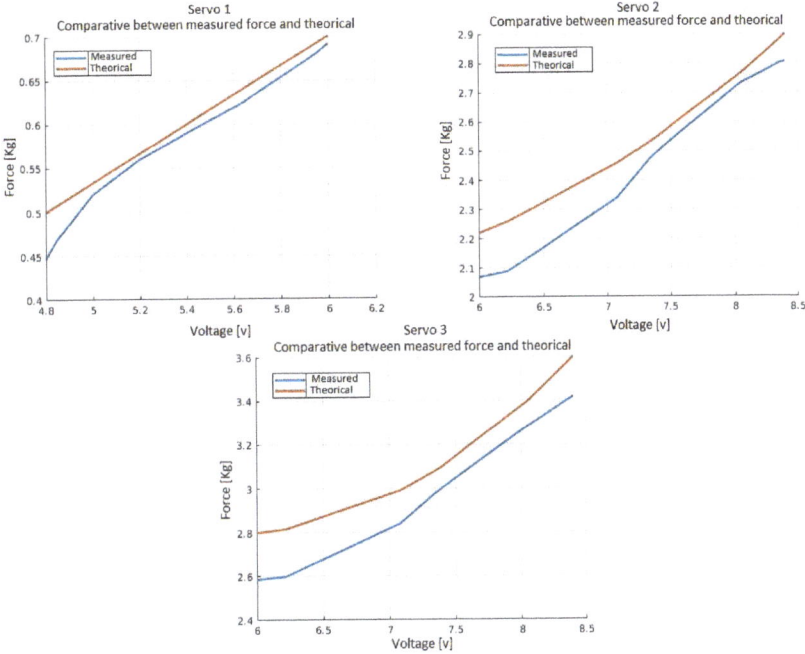

Figure 10. Comparison between the theoretical force and the real force performed by the soft limbs according to the voltage applied to the servos. The orange line represents the theoretical measurements and the blue line represents the experimental measurements obtained.

4.2. Contact Pressure

This section introduces the experiments for measuring the pressure exerted by the soft landing gear on two pipes of 140 mm and 160 mm diameter. In these experiments, force-sensing resistors (FRS sensors) have been distributed all over the soft surface. These sensors have a resistance that changes when a force is applied to it. This measurement can be mapped to forces and extrapolated to pressure over the surface. In these experiments, we tried to understand the behavior of the soft limbs and the pressure areas, where the soft train exerts less pressure on the pipes, which exert more pressure. Several experiments were carried out to calculate the pressure of each limb. Once the experiments were carried out, the data were collected and averaged to later be processed and obtain the pressure map. This process was done for both 160 mm and 140 mm pipes.

Figure 11 shows that more pressure is exerted in the base of the soft train and also in the tips of the limbs. The areas where less pressure is exerted are the intermediate areas due to formed folds.

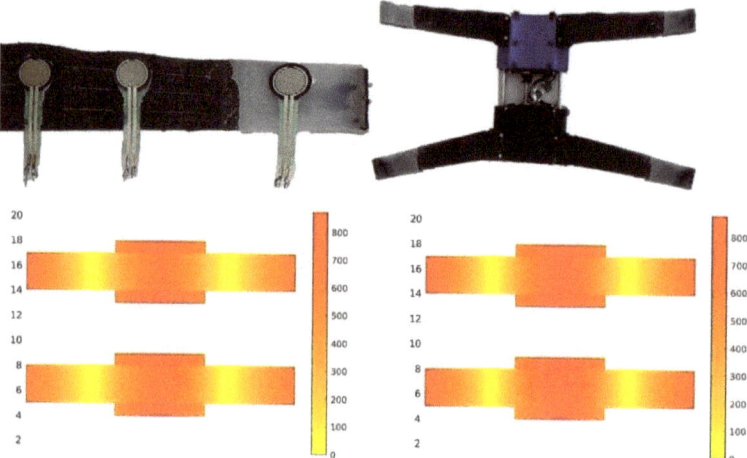

Figure 11. The upper image shows the lower part of the landing gear to which the pressure study is made. The lower left-hand image shows the pressure map made on a 140 mm-diameter pipe and the lower right-hand image shows the pressure map made on a 160 mm-diameter pipe.

The same behavior is observed in both studied cases, in the pressure graph made on the 160 mm-diameter pipe and in the 140 mm-diameter pipe. The difference between these two graphs is that the general pressure recorded on the 140 mm pipe pressure graph is lower; that is, in general, there is less pressure at the end of the extremities and at the base of the soft train than on the 160 mm pipe.

4.3. Maximum Lateral Angle

A study was also conducted to determine the range of the angle at which the soft train can be attached to the pipe without separating from it. For this test, a test-bench was installed in which a smooth PVC pipe of 160 mm diameter was placed and the maximum inclination angle concerning the vertical of the pipe was checked. The soft train together with the UAV was able to hold on to it with a maximum angle of 30 °C. Figure 12 shows the maximum angle at which the soft gripper can hold the contact to the pipe.

This experiment also tested the movement of the soft train and verified that the prominences arranged at the base of the train can pass over the pipe joints and their irregularities.

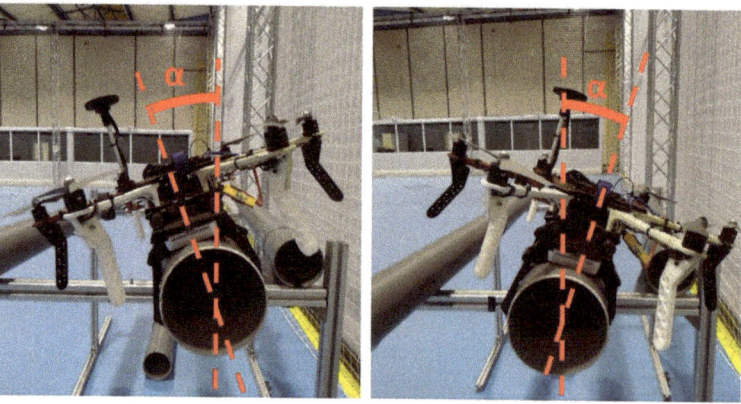

Figure 12. The maximum angle at which the landing gear can be grabbed with the drone on the pipe.

4.4. Crawling Gait Analysis

Finally, to analyze the repetitiveness of the movement of the system, a gait analysis using a motion capture (MoCap) system was carried out.

The MoCap system allowed us to record in real time the position of markers in a controlled movement. Three reflective markers have been placed at each limb, which are located at each joint where the limb is bent in Figure 13.

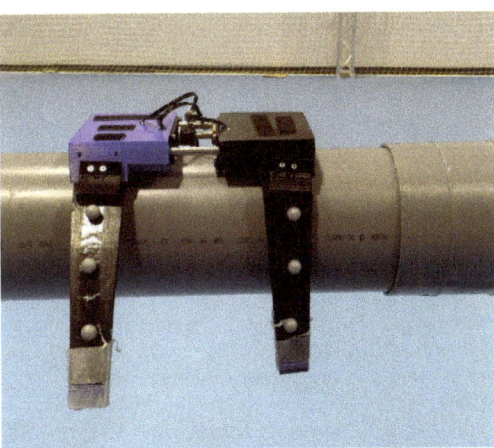

Figure 13. Placement of markers on a pair of soft limbs.

Knowledge of the position of the markers can be used to validate the motion on the pipeline. When the limbs are extended, the markers line up. As the limbs begin to bend, the markers move inward, forming a semicircle. Then, the displacement begins. First, the forward pair of limbs open and move forward. In the next step, the opposite happens: the front limbs are closed and the rear limbs open, which moves the mechanism forward. The motion must be linear on the pipe. Slippage is corrected by changing the center of gravity of the UAV by making adjustments when joining the soft landing gear to the main UAV platform to ensure the position of the center of gravity.

The recording of the position of the markers and the time can also be used to obtain the speed of the landing gear along the pipe. Thus, it has been obtained that the average speed is 4 cm/s.

Moreover, the position of the three markers on the limbs can also be used to obtain the radius of the circumference when closing the limbs. For the case studied, the radius is 84 mm.

Finally, it can be concluded that the movement made by the soft landing gear on the pipe is always the same, obtaining the same circumference radius. This also allows us to correct the center of gravity to prevent slipping, and to ensure that the UAV and the soft landing gear are centered on the pipe.

Figure 14 illustrates the motion of the marker points in two limbs.

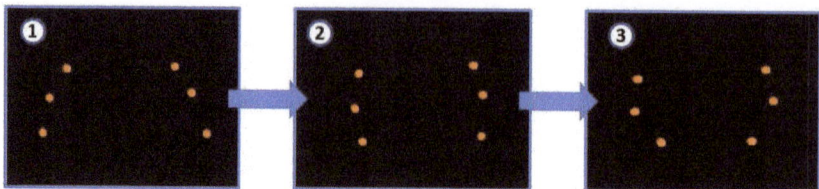

Figure 14. Closing sequence on the pipe and the limbs' deformation.In the first picture, the tentacles are open. In the second, they begin to close. Finally, in the third picture, the limbs are completely close, adapting to the shape of the pipe.

5. Experimental Test

This section describes the experimental setup of the whole working system, including the flying platform.

The soft landing gear was validated on a DJI FlameWhell 550 multirotor platform. This platform was chosen due to its versatile design. One of the advantages of the developed landing gear system is that it can be adapted to any type of multi-rotor system thanks to its modular design. It can also land on pipes of different diameters, and once landed, it can crawl along the pipe to perform both visual and contact inspection.

The motor controller board is in charge of supplying the necessary power for the motors to work and to control the motor sending the information through the data bus. An AVR-based board is used to control the landing and crawling on the pipe through the controller board, which executes the program that sequentially opens and closes the soft limbs to generate the movement of the system.

The AVR-based board is connected to an Nvidia tx2 onboard computer, which is in charge of sending the order received from the pilot and execute the high-level behavior software. However, this point is not the subject of this paper. It can also be used to integrate more sensors, including cameras. All these components are powered through specific voltage converters that regulate the power of each device from the battery.

The AVR-based board sends the information by serial communication to the motor's controller. This asynchronous Rx–Tx protocol has been used because it allows us to have two lines: one for transmitting to the motors and the other for receiving data such as the position, the speed, or which motor is working in each step. Figure 15 shows all of the components that we have used.

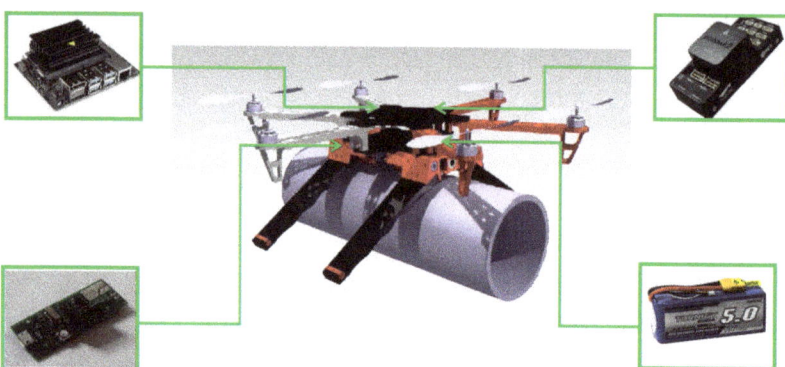

Figure 15. Setup including in the flying platform.

This UAV has two operation modes: manual and autonomous. In the manual mode, an operator sends the commands to move through the pipe. These commands are received by the Nvidia tx2 and transmitted to the AVR-based board via USB. The operator can select between either opening and closing the limbs, or either moving the soft train forward or backward. In the autonomous mode, the AVR-based board sends a sequential program to the motors so that they carry out a movement. A linear sequence of opening front limbs, moving forward, closing front limbs, opening rear limbs and moving forward is performed. This mode should be activated once the UAV is attached to the pipe.

Tests were carried out to verify the functionality of this unit. The first experiments that were performed on the pipe were the ones in which the soft landing gear moves through it. Once this stage was tested, the landing gear was installed on the multi-rotor system, in this case a DJI f550 which has a Pixhawk 2 for the control of the multi-rotor, an Nvidia tx2 as an onboard computer, a camera connected to it to locate and position the UAV on the pipe, and a gas sensor which is a safety sensor to detect a gas leak and preventing putting the installation at risk—which is further detailed in [32]. Figure 16 shows the scheme of the proposed system.

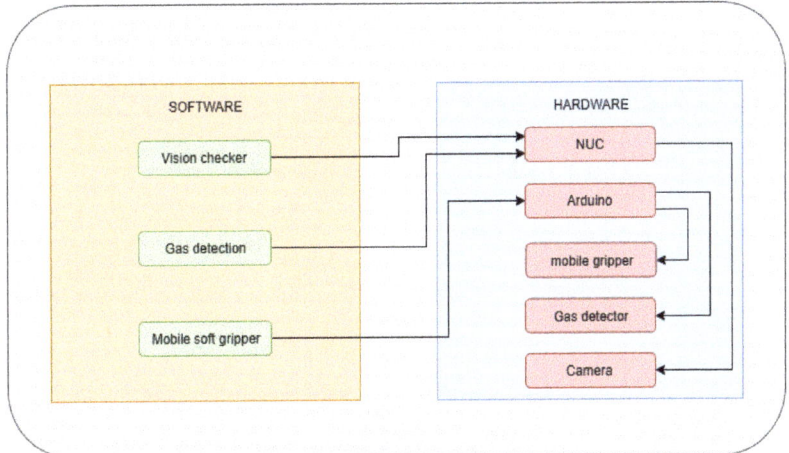

Figure 16. System scheme used.

Multiple experiments have been carried out to demonstrate that the system works under different conditions, including indoor and outdoor experiments with wind, which makes it more difficult to maneuver and land the UAV. The soft landing gear grips firmly to the pipe and helps center the multi-rotor to the pipe while landing. The grip is fast and safe and does not damage or scratch the pipe at any time, though the landing gear is strong enough to hold on to the pipe and move along it without needing to be stabilized with the UAV propellers. Figure 17 gives an example of the soft train attached to a pipe with and without the multi-rotor UAV.

Figure 17. The image on the left-hand side shows the first experiment performed with the landing gear alone on the pipe. The picture on the right-hand side shows the complete system and how it is attached to the pipe.

The gripper system worked in all of the tests, demonstrating its reliability and ability to overcome joints between pipes while moving. It can also be attached to irregular surfaces.

As proof of concept, an ultrasonic sensor was attached to the soft landing gear to acquire non-destructive information about the thickness of a typical steal pipe. Figure 18 shows the assembly and the components used for the inspection. The ultrasonic sensor is placed at the front of the landing gear and has a motor to move the sensor up and down. It is important to make the right pressure so it can take the measurements.

Figure 18. The image shows the complete system with an ultrasonic sensor and the flaw detection computer.

6. Conclusions

A novel design of a soft-tentacle gripper for UAVs to crawl on pipes was presented.

How the soft gripper was manufactured with 3D printing and molding techniques for the silicone-based material was also described. This manufacturing process is highly repeatable and reliable.

The design of the UAV landing gear is unique and is capable of crawling on pipes without damaging them. It also provides a fast coupling and decoupling of the pipe. It can move through welds and joints, without losing adhesion and without jamming. The soft mechanism also adapts to pipes of different diameters or even to non-cylindrical pipes.

The soft gripper was designed to be as lightweight as possible, easy to transport and easy to change. The system also has low-cost implementation and is easy to reproduce using additive manufacturing techniques.

The system has been validated with real tests landing and crawling on the pipes with an automatic system.

This system can be combined with various sensors to perform inspections, an example of which is the use of an ultrasonic sensor. This type of sensor combined with this soft landing gear is very useful for inspecting pipes located at high altitudes. This saves in terms of inspection costs and time and increases the safety of workers.

This research paper represents the first step to create a fully autonomous hybrid flying and crawling contact inspection robot that is able to operate in an environment with many obstacles.

Future work will focus on the development of a faster locomotion system with more degrees of freedom. Furthermore, a standard ultrasonic transducer will be introduced in the base of the crawl system in order to realize flaw detection over the pipe and to detect anomalies.

Author Contributions: Conceptualization, F.J.G.R., P.R.S. and B.C.A.; methodology, F.J.G.R., P.R.S. and B.C.A.; software, F.J.G.R. and P.R.S.; validation, F.J.G.R.; formal analysis, F.J.G.R.; investigation, F.J.G.R., P.R.S. and B.C.A.; resources, F.J.G.R., P.R.S. and B.C.A.; data curation, F.J.G.R.; writing–original draft preparation, F.J.G.R. and P.R.S.; rewriting—review and editing, F.J.G.R., P.R.S., and B.C.A.; visualization, F.J.G.R., P.R.S. and B.C.A.; supervision, P.R.S. and B.C.A.; project administration and supervision, A.O.; funding acquisition, A.O. All authors have read and agreed to the published version of the manuscript.

Funding: This work was funded by the European funded project HYFLIERS (Hybrid Flying–Rolling with-Snake-Arm Robot for Contact Inspection) SI-1762/23/2017 EU-funded project.

Institutional Review Board Statement: Not applicable.

Informed Consent Statement: Not applicable.

Data Availability Statement: Not applicable.

Acknowledgments: We thank Robotics, Vision and Control Group and GRIFFIN (General Compliant Aerial Robotic Manipulation System Integrating Fixed and Flapping Wings to Increase Range and Safety) SI-1867/23/2018 ERC-ADG EU-funded project for supporting us during this work. This work is part of the European Union-funded project HYFLIERS, which aims to develop an inspection and maintenance system for the oil and gas industry using hybrid (aerial and ground) robots (Hybrid Flying–Rolling with-Snake-Arm Robot for Contact Inspection) SI-1762/23/2017 EU-funded project; we also thank Alejandro Gomez Tamm and Vicente Perez Sanchez for their support of this paper.

Conflicts of Interest: The authors declare no conflict of interest.

Abbreviations

The following abbreviations are used in this manuscript:

UAV	Unmanned aerial vehicle
DOF	Degrees of freedom
DEA	Dielectric elastomer actuators
AM	Aerial manipulators
CAD	Computer-aided design
TPU	Thermoplastic polyurethane
PLA	Polylactic acid
ABS	Acrylonitrile butadiene styrene
EMF	Electromotive force
MoCap	Motion capture
NDT	Non-destructive testing

References

1. Ohta, P.; Valle, L.; King, J.; Low, K.; Yi, J.; Atkeson, C.G.; Park, Y. Design of a Lightweight Soft robotic Arm Using Pneumatic artificial Muscles and Inflatable sleeves. *Soft Robot.* **2018**, *5*, 204–215. [CrossRef] [PubMed]
2. Marchese, A.D.; Onal, C.D.; Rus, D. Autonomous-soft robotic fish capable of escape maneuvers using fluidic elastomer actuators. *Soft Robot.* **2014**, *1*, 75–87. [CrossRef] [PubMed]
3. Abeach, L.A.A.; Nefti-Meziani, S.; Davis, S. Design of avariable stiffness soft dexterous gripper. *Soft Robot.* **2017**, *4*, 274–284. [CrossRef] [PubMed]
4. Yap, H.K.; Ng, H.Y.; Yeow, C.H. High-force soft printable pneumatics for soft robotic applications. *Soft Robot.* **2016**, *3*, 144–158. [CrossRef]
5. Guo, J.; Xiang, C.; Helps, T.; Taghavi, M.; Rossiter, J. Elec-troactive textile actuators for wearable and soft robots. In Proceedings of the 2018 IEEE International Conference on Soft Robotics (Ro-boSoft), Livorno, Italy, 24–28 April 2018; pp. 339–343.
6. Khin, P.M.; Yap, H.K.; Ang, M.H.; Yeow, C. Fabric-based actuator modules for building soft pneumatic structures with high payload-to-weight ratio. In Proceedings of the 2017 IEEE/RSJ International Conference on Intelligent Robots and Systems (IROS), Vancouver, BC, Canada, 24–28 September 2017; pp. 2744–2750.
7. Low, J.H.; Cheng, N.; Khin, P.M.; Thakor, N.V.; Kukreja, S.L.; Ren, H.L.; Yeow, C.H. A bidirectional softpneu-matic fabric-based actuator for grasping applications. In Proceedings of the 2017 IEEE/RSJ International Conference on Intelligent Robots and Systems (IROS), Vancouver, BC, Canada, 24–28 September 2017; pp. 1180–1186.
8. Yan, J.; Zhang, X.; Xu, B.; Zhao, J. A new spiral-type inflatable pure torsional soft actuator. *Soft Robot.* **2018**, *5*, 527–540. [CrossRef] [PubMed]
9. Han, K.; Kim, N.-H.; Shin, D. A novel soft pneumatic artificial muscle with high-contraction ratio. *Soft Robot.* **2018**, *5*, 554–566. [CrossRef] [PubMed]
10. Mutlu, R.; Alici, G.; Panhuis, M.I.H.; Spinks, G.M. 3D printed flexure hinges for soft monolithic prosthetic fingers. *Soft Robot.* **2016**, *3*, 120–133. [CrossRef]
11. Marchese, A.D.; Katzschmann, R.K.; Rus, D. Arecipefor soft fluidic elastomer robots. *Soft Robot.* **2015**, *2*, 725.
12. Venter, D.; Dirven, S. Self morphing soft-robotic gripper for handling and manipulation of delicate produce in horticultural applications. In Proceedings of the 2017 24th International Conference on Mecha-tronics and Machine Vision in Practice (M2VIP), Auckland, New Zealand, 21–23 November 2017.
13. Sareh, P.; Chermprayong, P.; Emmanuelli, M.; Nadeem, H.; Kovac, M. Rotorigami: A rotary origami protective system for robotic rotorcraft. *Sci. Robot.* **2018**, *3*, 22. Available online: https://robotics.sciencemag.org/content/3/22/eaah5228 (accessed on 5 March 20021). [CrossRef] [PubMed]
14. Phan, H.V.; Park, H.C. Mechanisms of collision recovery in flying beetles and flapping-wing robots. *Science* **2020**, *370*, 1214–1219. Available online: https://science.sciencemag.org/content/370/6521/1214 (accessed on 27 March 2021). [CrossRef] [PubMed]
15. Chen, Y.; Zhao, H.; Mao, J.; Chirarattananon, P.; Helbling, E.F.; Hyun, N.-S.P.; Clarke, D.; Wood, R.J. Controlled flight of a microrobot powered by soft artificial muscles. *Nature* **2019**, *575*, 324–329. [CrossRef] [PubMed]
16. Mintchev, S.; Shintake, J.; Floreano, D. Bioinspired dual-stiffness origami. *Sci. Robot.* **2018**, *3*, eaau0275. Available online: https://robotics.sciencemag.org/content/3/20/eaau0275 (accessed on 20 March 2021). [CrossRef] [PubMed]
17. Nagase, J.; Fukunaga, F. Development of a novel crawler-mechanism for pipe inspection. In Proceedings of the in IECON 2016 42nd Annual Conference of the IEEE Industrial Electronics Society, Florence, Italy, 23–26 October 2016; pp. 5873–5878.
18. Kakogawa, A.; Ma, S. Design of an underactuated parallelogram crawler module for an in-pipe robot. In Proceedings of the 2013 IEEE International Conference on Robotics and Biomimetics (ROBIO), Shenzhen, China, 12–14 December 2013; pp. 1324–1329.
19. Schempf, H.; Mutschler, E.; Gavaert, A.; Skoptsov, G.; Crowley, A. Visual and non destructive evaluation inspection of live gas mains using the explorer family of pipe robots. *J. Field Robot.* **2010**, *27*, 217–249. [CrossRef]
20. Xu, X.; Wang, H.; Xie, F.; Wang, B. Study of the drive and speed governing control for a pipeline pig. In Proceedings of the 2018 10th International Conference on Modelling, Identification and Control (ICMIC), Guiyang, China, 2–4 July 2018; pp. 1–6.
21. Singh, P.; Ananthasuresh, G.K. A compact and compliant external pipe-crawling robot. *IEEE Trans. Robot.* **2013**, *29*, 251–260. [CrossRef]
22. Hausamann, D.; Zirnig, W.; Schreier, G.; Strobl, P. Monitoring of gas pipelines—A civil UAV application. *Aircr. Eng. Aerosp. Technol.* **2005**, *77*, 352–360. [CrossRef]
23. Bretschneider, T.; Shetti, K. Uav-based gas pipeline leak detection. In Proceedings of the 35th Asian Conference on Remote Sensing, ACRS2015: Sensing for Reintegration of Societies, Quezon, Manila, Philippines, 24–28 October 2015.
24. Gomez, C.; Green, D. Small unmanned airborne sys-tems tosupport oil and gas pipeline monitoring and mapping. *Arab. J. Geosci.* **2017**, *10*, 202. [CrossRef]
25. Ollero, A.; Siciliano, B. *Aerial Robotic Manipulation*; Springer: Berlin/Heidelberg, Germany, 2019.
26. Ollero, A.; Heredia, G.; Franchi, A.; Antonelli, G.; Kondak, K.; Sanfeliu, A.; Viguria, A.; Dios, J.R.M.; Pierri, F.; Cortes, J.; et al. The aeroarms project: Aerial robots with advanced manipulation capabilities for inspection and maintenance. *IEEE Robot. Autom. Mag.* **2018**, *25*, 1223. [CrossRef]
27. Mattarand, R.; Kalai, R. Development of a wall-sticking drone for non-destructive ultrasonic and corrosion testing. *Drones* **2018**, *2*, 8. [CrossRef]

28. Ramon-Soria, P.; Gomez-Tamm, A.; Garcia-Rubiales, F.; Arrue, B.; Ollero, A. Autonomous landing on pipes using soft gripper for inspection and maintenance in outdoor environments. In Proceedings of the 2019 IEEE International Conference on Robotics and Automation (ICRA), Macau, China, 3–8 November 2019.
29. Popek, K.M.; Johannes, M.S.; Wolfe, K.C.; Hegeman, R.A.; Hatch, J.M.; Moore, J.L.; Katyal, K.D.; Yeh, B.Y.; Bamberger, R.J. Autonomous grasping robotic aerial system for perching (agrasp). In Proceedings of the 2018 IEEE/RSJ International Conferenceon Intelligent Robots and Systems (IROS), Madrid, Spain, 1–5 October 2018; pp. 1–9.
30. Nguyen, H.; Siddall, R.; Stephens, B.; Navarro-Rubio, A.; Kova, A. A passively adaptive microspine grapple forrobust, controllable perching. In Proceedings of the 2019 2nd IEEE International Conference on Soft Robotics (Robo-Soft), Seoul, Korea, 14–18 April 2019; pp. 80–87.
31. Tieu, M.; Michael, D.; Pflueger, J.; Sethi, M.; Shimazu, K.; Anthony, T.; Lee, C. Demonstrations of bio-inspired perching landing gear for UAVs. In Proceedings of the Bioinspiration, Biomimetics, and Bioreplication 2016, International Society for Optics and Photonics, Las Vegas, NV, USA, 20–24 March 2016.
32. Garcia-Rubiales, F.J.; Ramon-Soria, P.; Arrue, B.C.; Ollero, A. Magnetic detaching system for modular UAVs with perching capabilities in industrial environments. In Proceedings of the 2019 Workshop on Research, Education and Development of Unmanned Aerial Systems (RED UAS), Cranfield, UK, 25–27 November 2019; pp. 172–176.
33. Doyle, C.E.; Bird, J.J.; Isom, T.A.; Kallman, J.C.; Bareiss, D.F.; Dunlop, D.J.; King, R.J.; Abbott, J.J.; Minor, M.A. An avian-inspired passive mechanism for quadrotorperching. *IEEE/ASME Trans. Mech.* **2013**, *18*, 506–517. [CrossRef]

Detection of the Deep-Sea Plankton Community in Marine Ecosystem with Underwater Robotic Platform

Jiaxing Wang [1], Mingqiang Yang [1,2,*], Zhongjun Ding [3,*], Qinghe Zheng [1], Deqiang Wang [1], Kidiyo Kpalma [4] and Jinchang Ren [5]

1 School of Information Science and Engineering, Shandong University, Jinan 266237, China; 17854264303@163.com (J.W.); 15005414319@163.com (Q.Z.); wdq_sdu@sdu.edu.cn (D.W.)
2 Shenzhen Research Institute, Shandong University, Shenzhen 518000, China
3 China National Deep Sea Center, Qingdao 266237, China
4 IETR-INSA UMR-CNRS 6164, 35708 Rennes, France; kidiyo.kpalma@insa-rennes.fr
5 Centre for Excellence in Signal and Image Processing, Department of Electronic and Electrical Engineering, University of Strathclyde, Glasgow G1 1XQ, UK; jinchang.ren@strath.ac.uk
* Correspondence: yangmq@sdu.edu.cn (M.Y.); dzj@ndsc.org.cn (Z.D.)

Abstract: Variations in the quantity of plankton impact the entire marine ecosystem. It is of great significance to accurately assess the dynamic evolution of the plankton for monitoring the marine environment and global climate change. In this paper, a novel method is introduced for deep-sea plankton community detection in marine ecosystem using an underwater robotic platform. The videos were sampled at a distance of 1.5 m from the ocean floor, with a focal length of 1.5–2.5 m. The optical flow field is used to detect plankton community. We showed that for each of the moving plankton that do not overlap in space in two consecutive video frames, the time gradient of the spatial position of the plankton are opposite to each other in two consecutive optical flow fields. Further, the lateral and vertical gradients have the same value and orientation in two consecutive optical flow fields. Accordingly, moving plankton can be accurately detected under the complex dynamic background in the deep-sea environment. Experimental comparison with manual ground-truth fully validated the efficacy of the proposed methodology, which outperforms six state-of-the-art approaches.

Keywords: image motion analysis; image processing; optical flow; underwater robotic

1. Introduction

Plankton are organisms that live in oceans and fresh water [1] that play an important role in the material and energy recycling within the marine food chain [2]. The study of plankton community and plankton itself is indispensable for understanding of marine resources and the impacts of climate change on ecosystems [3]. In addition, the number of plankton is a key indicator of carbon and energy cycling [4], and of great significance to species diversity and ecosystem diversity [5]. From the early 19th century to date, many examples of large-scale sensor equipment were used to solve the challenge of getting reliable high-resolution estimates of plankton abundance at depth [6]. Acoustic and optical techniques for the in-situ observation of zooplankton are currently popularly used for plankton distribution assessment. Although acoustic-based observation has outstanding advantages of high observation frequency, it has inaccurate quantification and usually requires the combination of optical image analysis or other traditional sampling of zooplankton. In recent years, a series of advances were made in computer vision [7], including hyperspectral imaging [8], principal component analysis of images [9,10], and deep learning [11–13] for image classification [14]. As marine plankton is small and uneven in size, it is difficult to describe it quantitatively, such as with inventory and abundance statistics.

At present, a lot of plankton detection methods are proposed that often rely heavily on the use of sophisticated underwater instruments. J. Craig et al. [15,16] constructed an ICDeep system, based on the Image Intensified Charge Coupled Device (ICCD) camera, to assess the quantity of low-light bioluminescent sources in the marine environment. Philips et al. [17] created a marine biological detector, where a Scientific CMOS (SCMOS) camera was used to image the organisms before conducting statistical analysis of the plankton abundance. With the development of the computer vision, multitarget tracking-enabled automatic analysis was gradually applied to this field [18]. Kocak et al. [19] proposed to use the active contour (snake) models to segment, label, and track images of the snake model for the classification of the plankton. Luca et al. [20] also presented an automatic plankton counting method, which mainly used the interframe difference and the intersection of the bounding boxes to perform multitarget matching. The aforementioned methods achieved some results in automatic analysis and counting. However, there are still some challenges due to the particularity and complexity of plankton's own form and passive movement mode. Applying machine vision techniques to underwater images or videos is a feasible way to study plankton at present. Underwater plankton imaging has the capacity to detect patterns of the plankton distributions that we would be unable to be tackled by sampling with nets. [21]. Therefore, we consider applying machine vision technology to underwater images or videos is currently a feasible method for studying plankton.

Underwater robots play an important role in various video surveillance tasks including data collection. A mobile robot that can be fixed on a rotatable axis would be advantageous because it provides 360° visual coverage instead of using a fixed image camera installed in a predetermined direction. These mobile robots capture unprecedented shots of marine life in dangerous environments inaccessible to humans. A submarine can push and control the underwater robot to complete the collection of deep-sea data and store the data in the computer for analysis. Some underwater robots are shown in Figure 1.

In this paper, we propose a deep-sea plankton detection method based on the Horn–Schunck (HS) optical flow [22]. The optical flow is the instantaneous velocity of the pixel movement of the moving object on the image plane. The advantage of the optical flow method is that the motion vectors can be estimated by the optical flow vector accurately. In this way, one can detect the plankton and easily analyze statistically its volume using image processing and machine vision. The research on plankton can be specifically divided into density, position, number, individual and total volume, etc. In the case where the spatial position of plankton does not coincide in two consecutive frames, the presence or absence of plankton should be determined according to the following conditions: the time gradient maps at the plankton's location in two consecutive optical flow fields will be opposite to each other, and the horizontal and vertical gradients of the plankton at that location are equal and their direction is the same. Since the connected components are marked as the location of plankton, the number of connected components can be regarded as the number of plankton. By using this method, we firstly count the number of plankton in the video, followed by a statistical analysis. Various comparative experiments are carried out to benchmark with other methods to fully demonstrate the effectiveness of the proposed methodology.

Figure 1. Underwater robot pattern: submarine can push and control underwater robot to complete collection of deep-sea data, then store data in computer for analysis.

2. The Proposed Method

2.1. Principle

The deep ocean floor is clear and suitable for video acquisition with active lighting. During the video acquisition process, the camera position and shooting angle change with the movement of the submersible, making the plankton detection task a moving target detection problem under complex and dynamic backgrounds. Two consecutive optical flow field matrices derived from three consecutive video frames in a video are employed. For fast-moving plankton (plankton does not overlap in space in two consecutive frames), the two consecutive optical flow values at the position where the plankton is located are opposite. In practice, the amount of grayscale change is often close to 0. Therefore, the two consecutive optical flows are approximately opposite to each other, and we discuss this situation by setting two thresholds in the experiment section. We use this property to map out the location of the plankton. Figure 2, hereafter provides an overview of the proposed method, which consists of three modules.

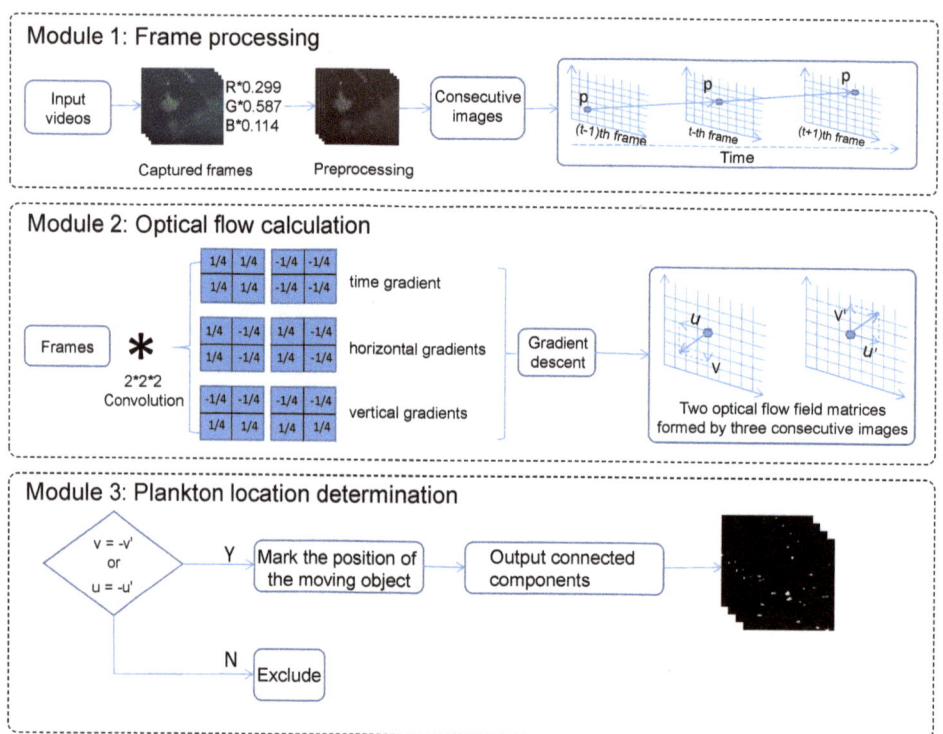

Figure 2. Main processing blocks of proposed algorithm. Module 1 is for preprocessing, whilst Module 2 performs 3D convolution on video frame to extract dense optical flow. Module 3 is a dual threshold setting to determine whether a plankton is contained at a specific location or not(see Section 3.3).

As shown in the Module 1 of Figure 2, grayscale images are obtained by weighting three channels of the input frames. In module 2, three convolution operations are performed on two consecutive frames to produce three different gradients(see Figure 3), which correspond to three different convolution kernels. The details of the convolution process are shown in Figure 3 to illustrate this process. We find that the time gradients of the two optical flow fields derived from three consecutive frames of images are opposite in numerical value and direction in the corresponding positions of plankton in the middle frame. In the following description, the time gradients of the two consecutive optical flow fields are represented by ∇_t and ∇'_t. The horizontal gradients of the two consecutive optical flow fields derived from three consecutive frames are equal in magnitude and direction in the corresponding positions of plankton in the middle frame. Similarly, the vertical gradients are also equal. In the following description, the horizontal gradients of the two optical flow fields are represented by ∇_x and ∇'_x, the vertical gradients are ∇_y and ∇'_y. Finally, Module 3 is for dual thresholding, which is explained separately when discussing the parameters later.

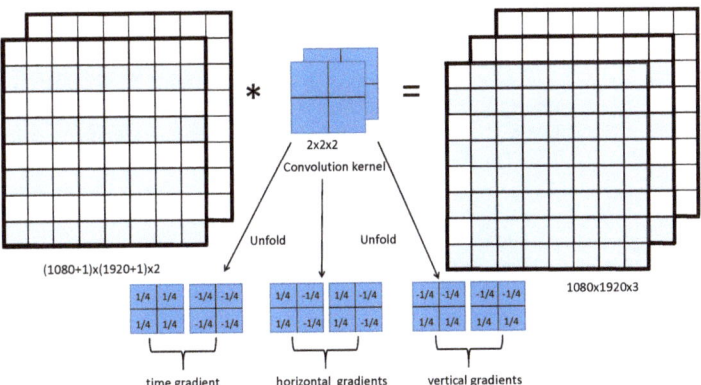

Figure 3. Three convolution kernels corresponding in time and space. Two consecutive frames are used to form a 3D matrix whose size is $(height + 1) \times (width + 1) \times 2$. Size of filter is $2 \times 2 \times 2$. Result of each operation is gradient of the pixel at upper-left corner of convolution kernel.

2.2. Proof

In the HS optical flow method, the constraint equation of optical flow can be established as Equation (2) according to the premise of the optical flow method: invariance of gray level [22]. Three first-order differences are used to replace the horizontal, vertical, and time gradients. Let the gray value at plankton's position in the middle frame be $I_{x,y,t}$, where the subscripts x and y are the pixel index, and t is the time index. The position of plankton changes with the movement of ocean current and the camera lens. As shown in Figure 4, the plankton is small-sized, so its position in frame t doesn't overlap in frame $t + 1$. When it changes from position 1 to position 2, the gray value corresponding to position 2 of plankton at frame $t - 1$ is the background gray value $I_{x,y,t-1}$. In a similar way, when the position of plankton changes from position 2 to position 3, the gray value corresponding to position 2 at frame $t + 1$ becomes the background gray value $I_{x,y,t+1}$. Based on the characteristics of deep-sea underwater video, the background around the plankton is invariant in time, i.e.,:

$$I_{x,y,t-1} = I_{x,y,t+1} \tag{1}$$

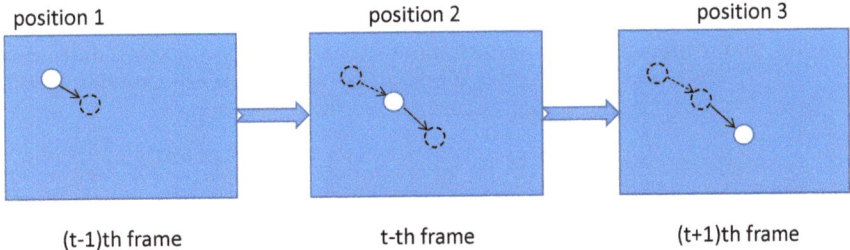

Figure 4. Position of plankton in three consecutive frames.

$$\nabla_x u + \nabla_y v + \nabla_t = 0 \tag{2}$$

The time gradients at the plankton's positions in the two adjacent optical flow fields are:

$$\nabla_t = \frac{1}{2}(I_{x,y,t} - I_{x,y,t-1} + I_{x+1,y,t} - I_{x+1,y,t-1}) \tag{3}$$

$$\nabla'_t = \frac{1}{2}(I_{x,y,t+1} - I_{x,y,t} + I_{x+1,y,t+1} - I_{x+1,y,t}) \tag{4}$$

Based on Equation (1), the background gray value $I_{x,y,t-1} = I_{x,y,t+1}$, $\nabla_t = -\nabla'_t$, the time gradients of the two optical flow fields derived from three consecutive frames of images are opposite in the corresponding positions of plankton in the middle frame.

The horizontal gradients of the plankton's location in the two optical flow fields are:

$$\nabla_x = \frac{1}{2}(I_{x+1,y,t} - I_{x,y,t} + I_{x+1,y,t-1} - I_{x,y,t-1}) \tag{5}$$

$$\nabla'_x = \frac{1}{2}(I_{x+1,y,t+1} - I_{x,y,t+1} + I_{x+1,y,t} - I_{x,y,t}) \tag{6}$$

The same way, based on Equation (1), we can get that $\nabla_x = \nabla'_x$, i.e., the horizontal gradients of the two optical flow fields derived from three consecutive frames are equal in the corresponding positions of plankton in the middle frame. In the same way, we can get $\nabla_y = \nabla'_y$.

In fact, in the process of proof, the time and space gradients are estimated in a $2 \times 2 \times 2$ cubic neighborhood by taking the mean.

Then, we iterate n times for gray gradient relaxation by setting the initial conditions as $v_0 = v'_0 = 0$ and $u_0 = u'_0 = 0$.

$$\Delta = \left(\frac{\nabla_x u_n + \nabla_y v_n + \nabla_t}{\alpha_2 + \nabla_x^2 + \nabla_y^2}\right) \tag{7}$$

$$u_{n+1} = u_n - \nabla_x \Delta \tag{8}$$

$$v_{n+1} = v_n - \nabla_y \Delta \tag{9}$$

The parameter α_2 reflects the smoothness constraints of the HS optical flow algorithm; Δ is an iteration factor in the process of the iterative algorithm; ∇_x and ∇_y are the horizontal and vertical gradients, and u and v are the horizontal and vertical optical flow field matrices, respectively.

The relationships of Equations (7)–(9) are represented by a series, where the number of iterations is n. Let's substitute Equations (7)–(9), the new formulas are as follows:

$$u_{n+1} = u_n - \nabla_x \left(\frac{\nabla_x u_n + \nabla_y v_n + \nabla_t}{\alpha_2 + \nabla_x^2 + \nabla_y^2}\right) \tag{10}$$

$$v_{n+1} = v_n - \nabla_y \left(\frac{\nabla_x u_n + \nabla_y v_n + \nabla_t}{\alpha_2 + \nabla_x^2 + \nabla_y^2}\right) \tag{11}$$

where u_{n+1} and u_n are two horizontal optical flow fields before and after the n-th iteration, v_{n+1} and v_n are two vertical optical flow fields before and after the n-th iteration. We can derive $u_{n+1} = -u'_{n+1}$, $v_{n+1} = -v'_{n+1}$. When $n = 0$, we have:

$$v_1 = v_0 - \nabla_y \left(\frac{\nabla_x u_0 + \nabla_y v_0 + \nabla_t}{\alpha_2 + \nabla_x^2 + \nabla_y^2}\right) \tag{12}$$

$$v'_1 = v'_0 - \nabla'_y \left(\frac{\nabla'_x u'_0 + \nabla'_y v'_0 + \nabla'_t}{\alpha_2 + \nabla'^2_x + \nabla'^2_y}\right) \tag{13}$$

v_1 and v'_1 are the two consecutive vertical optical flow field at the first iteration. If the time gradients of the last two optical flow fields are opposite, that is $\nabla_t = -\nabla'_t$, we can get: $v_1 = -v'_1$. When $n = k$, $v_{k+1} = -v'_{k+1}$. That is, Equations (14) and (15) are opposite:

$$v_{k+1} = v_k - \nabla_y \left(\frac{\nabla_x u_k + \nabla_y v_k + \nabla_t}{\alpha_2 + \nabla_x^2 + \nabla_y^2}\right) \tag{14}$$

$$v'_{k+1} = v'_k - \nabla'_y\left(\frac{\nabla_x u'_k + \nabla'_y v'_k + \nabla'_t}{\alpha_2 + \nabla'^2_x + \nabla'^2_y}\right) \tag{15}$$

where v_{k+1} and v'_{k+1} represent the previous and the next vertical optical flow field matrix at the $(k+1)$th iteration, respectively.

When $n = k + 1$, we can show that $v_{k+2} = -v'_{k+2}$

$$v_{k+2} = v_{k+1} - \nabla_y\left(\frac{\nabla_x u_{k+1} + \nabla_y v_{k+1} + \nabla_t}{\alpha_2 + \nabla^2_x + \nabla^2_y}\right) \tag{16}$$

$$v'_{k+2} = v'_{k+1} - \nabla'_y\left(\frac{\nabla'_x u'_{k+1} + \nabla'_y v'_{k+1} + \nabla'_t}{\alpha_2 + \nabla'^2_x + \nabla'^2_y}\right) \tag{17}$$

By adding Equations (16) and (17), and substituting $v_{k+1} = -v'_{k+1}$, $\nabla_x = \nabla'_x$ and $\nabla_y = \nabla'_y$ into Equation (16) and Equation (17), respectively, we have:

$$v_{k+2} = -v'_{k+2} \tag{18}$$

Therefore, for fast-moving plankton, the values of the vertical optical flow field matrices of the space position where the plankton is located are opposite from each other: $v = -v'$, and the same applies horizontally: $u = -u'$.

2.3. The Volume of Plankton

Based on the above proof, one can calculate the number of pixels where plankton is located, and then multiply the actual size of a pixel to obtain the area of plankton. The resolution of the known image is *height* × *width*. According to camera internal reference, the actual range of our field of view is about W m by H m. The calculation of the actual area is given by:

$$S = N \times (W/width) \times (H/height) \tag{19}$$

where N is the number of pixels, and S is the corresponding actual surface. A method of approximate calculation is adopted here. Firstly, we can get the radius of a circle that has the same area as the plankton, and then calculate the volume of the sphere based on that radius. The advantage of this method is that we can get the 3D volume of an irregular object only by its area [23]. In addition, we can predict the type of plankton based on the estimated size, laying the foundation for the later identification of plankton types. The volume can be calculated by:

$$V = \frac{4}{3} \times \pi^{-\frac{1}{2}} \times S^{\frac{3}{2}} \tag{20}$$

The proposed method adds its own theoretical innovation on the basis of the original optical flow method and was proved mathematically. In this way, the complexity and passive motion patterns of plankton are well-solved, and the accuracy improves as the above problems are solved.

3. Experimental Results and Analysis

The data capture was provided by the China National Deep Sea Center. The data set was obtained by an underwater robotic nondestructive testing system carried by a deep-sea manned submersible. The camera's technical specifications are: resolution: 1080*i* HDTV; minimum illumination: 2l ux; optical zoom: 10 times; digital zoom: 12 times; aperture range: 3.2 mm–32 mm; video aspect ratio: 16:9 or 4:3. In this study, three six-minute videos of the plankton community from appearing to disappearing from the screen were selected, which were obtained from a submarine on the western Pacific sea mountain slope, and the diving depths are 2741.88 m and 5555.68 m, corresponding to 76 and 77 dives, respectively. The reason why the three videos are selected is that plankton appeared more frequently in them. Due to the complexity of the deep-sea environment and the irregular camera movement, the background is complex and dynamic. In this case, using high-precision image processing technology to study the plankton community from appearing to disappearing from the screen can effectively distinguish sedimentary clouds and plankton community in images. Examples of deep-sea plankton images are shown in Figure 5 and the details of data set including diving number, date, diving time, longitude, latitude and depth are shown in Table 1.

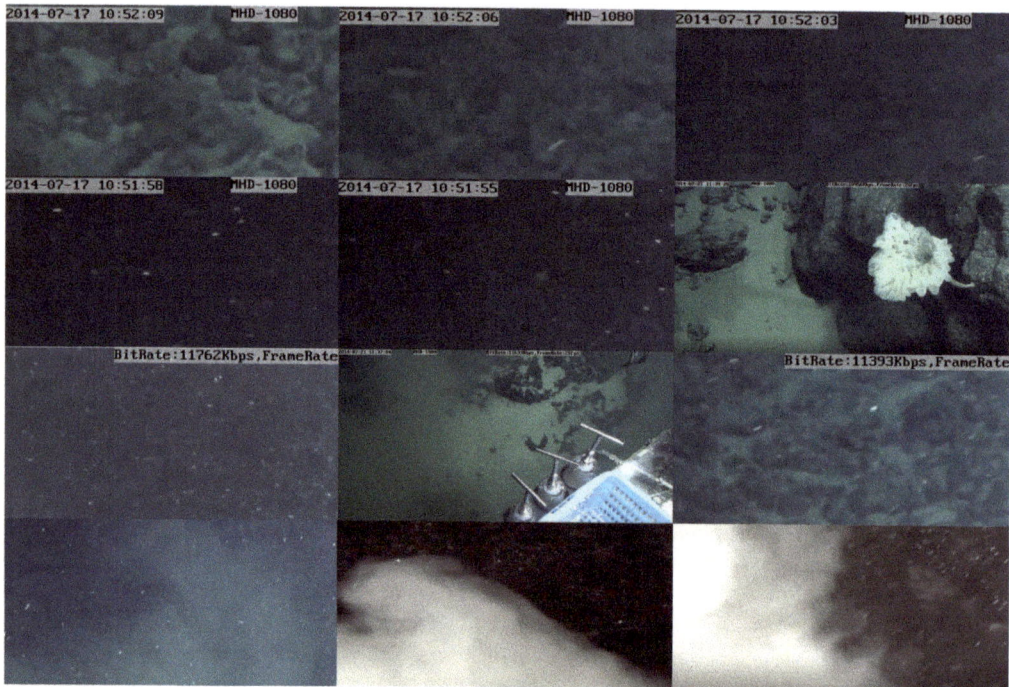

Figure 5. Example images of deep sea plankton.

Table 1. Details of datasets including diving number, date, diving time, longitude, latitude, and depth.

Diving Number	Date	Diving Time	Longitude	Latitude	Depth
76	17 July 2014	8.95 h	155.32° E–155.34° E	15.50° N–15.52° N	2741.88 m
77	21 July 2014	10.33 h	154.58° E–154.59° E	15.70° N–15.72° N	5555.68 m

3.1. Number and Volume of Plankton

Processing the recorded video of a complete plankton community from appearing to disappearing from the screen, the results obtained are shown in Figure 6. Figure 6a shows the variation of the number of plankton in three six-minute videos, and Figure 6b shows the variation of the volume of the corresponding three videos. The process of plankton appearing in front of the camera to disappearing is shown in Figure 6c,d. In the first 30 s of Figure 6c, the amount of plankton is small and the detection results are more accurate. We can see that the amount of plankton rises in the last 30 s of Figure 6c. For dense particle clouds, overlap, and hence, occlusion occurs frequently, which leads to relatively low average accuracy and recall rates.

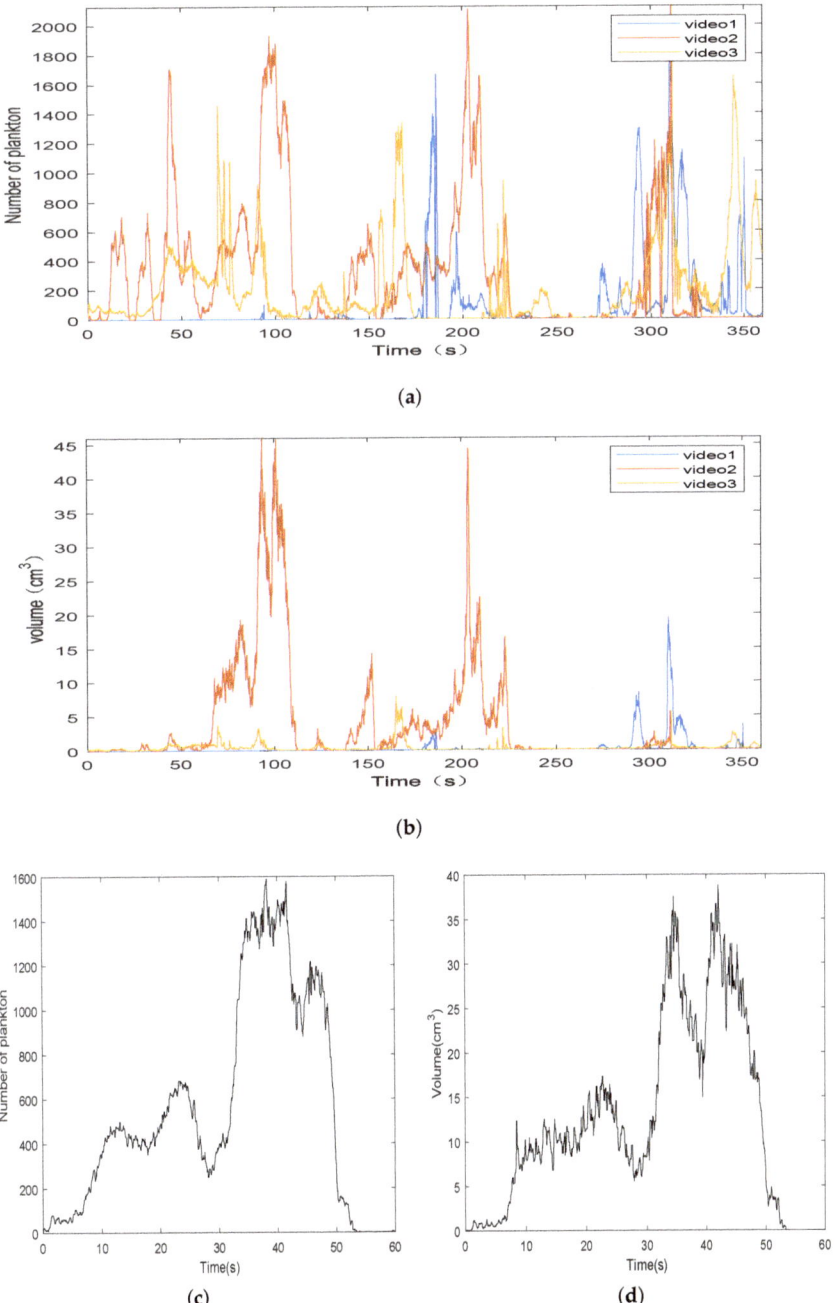

Figure 6. (**a**) Number of plankton in three six-minute videos. (**b**) Total volume of plankton in three six-minute videos. (**c**) Number of plankton in a period. (**d**) Volume of plankton in a period.

The actual volume curve of plankton in the video is shown in Figure 6b,d. We can see that the volume curve and the quantity curve of plankton generally follow the same trend. At the 40th second in Figure 6c, the plankton community moves away from the camera and then comes back, resulting in a smaller scene and a smaller overall volume due to perspective. So, we can see that the volume curve goes down and then goes up from Figure 6d.

3.2. Comparison with Six Target Detection Methods

The proposed method is compared with six state-of-the-art methods for performance evaluation. The results are shown in Figure 7, where Figure 7a represents some original images of the video, including sediment clouds, plankton, and uneven backgrounds. Top-Hat transform [24] is used to detect the location of the plankton in the image as shown in Figure 7b, the weakness of this algorithm is that there are some missed cases. Figure 7c and Figure 7d show the detection results of the frame difference method [25] and the motion estimation and image matching method [26], respectively. We show the result from the scan line marking method [27] in Figure 7e results from the simple block-based sum of absolute differences flow (SD) method [28], and the Lucas–Kanade (LK) optical flow method [29] are given in Figure 7f,g. The weakness of the above three methods is that there are a few false positives, and both Figure 7c,e detected the sediment cloud in the background by mistake. The result of Figure 7h is obtained using the proposed method. After comparing with the manual ground truth, we find that the plankton detected by the proposed method is more consistent with the original image in Figure 7a.

We take 20 images of the video, and the data are cleaned by manual counting to get the ground-truth. Then, we compare the number of plankton, recall rate, precision rate, and *F1-score* of the seven methods. When using 10 frames in the first 30 s of the video, the amount of plankton is small and the detection results are more accurate, the average accuracy rate is 0.901, the average recall rate is 0.955, and *F1-score* is 0.927. In addition, the equations and related symbols are shown in Table 2 and Equations (21)–(23). The results are shown in Tables 3 and 4. Taking 10 frames in the last 30 s of the video, the amount of plankton is high. For dense particle clouds, overlap can easily occur, and hence, occlusion occurs frequently, so the average accuracy and recall rates are relatively low, i.e., 0.895 and 0.943, respectively, and the *F1-score* is 0.918, The results are shown in Tables 5 and 6. In addition, we randomly selected 10 frames from the video for testing. The experimental results are shown in Tables 7 and 8. The performance of the proposed method is still very good. We use bold font to highlight the best results in each category in Tables 4, 6, 8 and 9.

$$Precision = \frac{TP}{TP + FP} \qquad (21)$$

$$Recall = \frac{TP}{TP + FN} \qquad (22)$$

$$F1 = \frac{2 Precision \times Recall}{Precision + Recall} \qquad (23)$$

Table 2. Confusion Matrix.

	Relevant	Nonrelevant
Retrieved	True Positives (TP)	False Positives (FP)
Not Retrieved	False Negatives (FN)	True Negatives (TN)

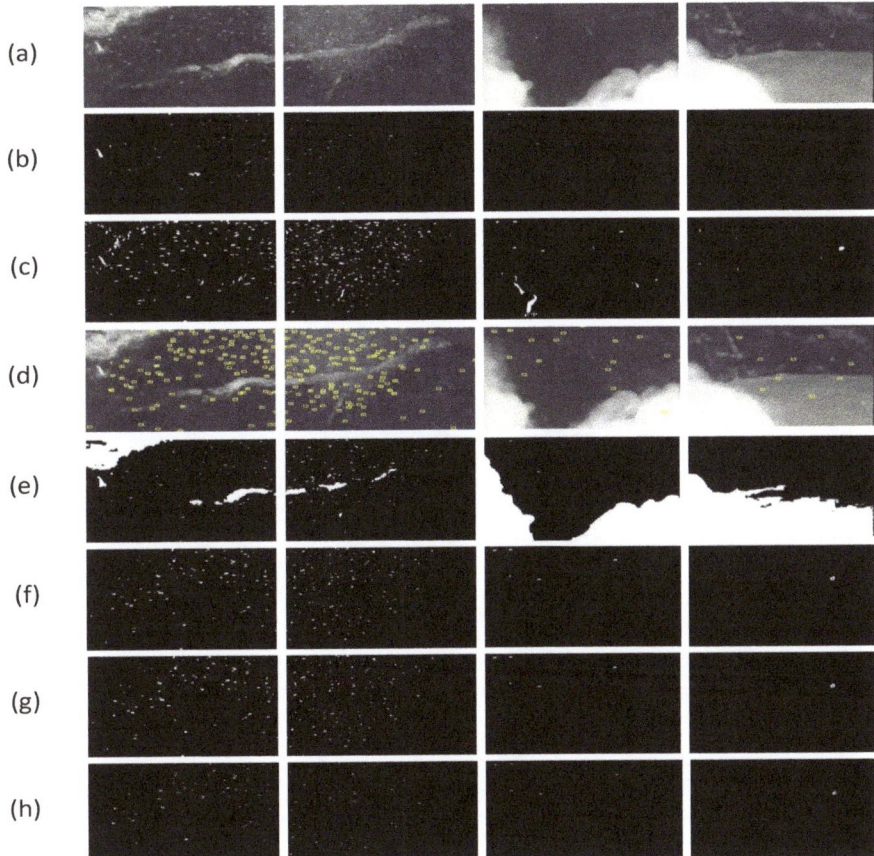

Figure 7. Location of plankton detected with seven different methods: (**a**) original image; (**b**) Top-Hat transform; (**c**) frame difference method; (**d**) motion estimation and image match; (**e**) scan line marking method; (**f**) simple block-based sum of absolute differences flow (SD); (**g**) Lucas–Kanade (LK) optical flow method, and (**h**) proposed method.

Table 3. In first 30 s, comparison of number of detected plankton using seven methods and Ground-Truth.

The Ten Frames:	1	2	3	4	5	6	7	8	9	10	Mean	std
Top-Hat	10	12	9	15	17	18	18	22	24	22	16.7	4.9
Frame difference	20	22	18	22	21	17	19	15	17	14	18.5	2.7
Image match	26	24	23	21	24	17	15	15	16	16	19.7	4.1
Scan line marking	11	7	8	9	10	9	10	9	9	8	9.0	1.1
SD	17	13	11	11	13	13	12	10	12	10	12.2	1.9
LK	17	14	13	13	13	12	11	11	11	10	12.7	1.8
Proposed method	16	14	12	12	12	11	12	10	11	10	12.0	1.7
Ground-Truth	14	13	12	12	11	12	11	9	10	9	11.3	1.6

Table 4. In first 30 s, comparison of recall rate, precision rate, and *F1-score* of seven methods.

	The Ten Frames:	1	2	3	4	5	6	7	8	9	10	Average
Top-Hat	Precision	0.9	0.85	0.89	0.73	0.59	0.61	0.56	0.36	0.42	0.41	0.632
	Recall	0.64	0.92	0.67	0.92	0.91	0.92	0.91	0.89	1	1	0.878
	F1	0.75	0.88	0.76	0.81	0.72	0.73	0.69	0.51	0.59	0.58	0.73
Frame difference	Precision	0.65	0.55	0.61	0.5	0.52	0.65	0.56	0.53	0.59	0.57	0.573
	Recall	0.93	0.92	0.92	0.92	1	0.92	1	0.89	1	0.89	0.939
	F1	0.77	0.69	0.73	0.65	0.68	0.76	0.72	0.66	0.74	0.69	0.712
Image match	Precision	0.54	0.54	0.49	0.52	0.46	0.65	0.67	0.53	0.56	0.56	0.552
	Recall	1	1	0.92	0.92	1	0.92	0.91	0.89	0.9	1	0.946
	F1	0.7	0.7	0.64	0.66	0.63	0.76	0.77	0.66	0.69	0.72	0.7
Scan line marking	Precision	0.82	0.86	0.88	0.78	0.9	0.89	0.9	0.89	0.89	0.88	0.869
	Recall	0.64	0.46	58	0.58	0.82	0.67	0.82	0.89	0.8	0.78	0.704
	F1	0.72	0.6	0.7	0.67	0.86	0.76	0.86	**0.89**	0.84	0.83	0.778
SD	Precision	0.76	0.92	0.91	0.91	0.77	0.85	0.83	0.8	0.75	0.8	0.83
	Recall	0.93	0.92	0.83	0.83	0.91	0.92	0.91	0.89	0.9	0.89	0.893
	F1	0.84	0.92	0.87	0.87	0.83	0.88	**0.87**	0.84	0.82	0.84	0.86
LK	Precision	0.76	0.86	0.85	0.85	0.85	0.85	0.83	0.73	0.82	0.8	0.82
	Recall	0.93	0.93	0.92	0.92	1	0.92	0.91	0.89	0.9	0.89	0.921
	F1	0.84	0.89	0.88	0.88	0.92	0.88	**0.87**	0.8	**0.86**	0.84	0.868
Proposed method	Precision	0.81	0.93	1	1	0.92	1	0.83	0.8	0.82	0.9	0.901
	Recall	0.93	1	1	1	1	0.92	0.91	0.89	0.9	1	0.955
	F1	**0.87**	**0.96**	**1**	**1**	**0.96**	**0.96**	**0.87**	0.84	**0.86**	**0.95**	**0.927**

Table 5. In last 30 s, comparison of number of detected plankton using seven methods and Ground-Truth.

The Ten Frames:	1	2	3	4	5	6	7	8	9	10	Mean	std
Top-Hat	22	17	17	16	13	12	13	12	9	18	14.9	3.6
Frame difference	28	28	30	24	30	15	28	28	27	24	26.2	4.2
Image match	22	22	25	26	31	27	32	28	31	24	26.8	3.5
Scan line marking	13	15	14	11	16	15	19	15	15	13	14.6	2.0
SD	16	21	22	23	23	23	22	23	20	17	21.0	2.4
LK	16	22	23	22	23	23	21	23	20	18	21.1	2.4
Proposed method	15	21	22	21	22	21	21	22	19	16	20.0	2.4
Ground-Truth	19	19	19	18	21	18	21	21	18	15	18.9	1.8

3.3. Discussion of Parameters

For each imaging system, there is a depth of field within which the closest field objects and farthest field objects are all in focus. If we deploy the system in air, the light intensity for the near field object and far field object should not be different in theory. However, when deployed in seawater, the light intensity changes as the light propagates in the water from near-field to far-field because of scattering caused by seawater and particles in the seawater. Therefore, during the experiment, there are two situations that need to be discussed. Firstly, 'grayscale invariance' is one of the prerequisites of the HS optical flow method, but in actual operation, the amount of grayscale change is often close to 0 but not equal to 0. Therefore, the threshold β_1 is set to handle this situation, as shown in Equation (24).

Table 6. In last 30 s, comparison of recall rate, precision rate, and F1-score of seven methods.

	The Ten Frames:	1	2	3	4	5	6	7	8	9	10	Average
Top-Hat	Precision	0.77	0.94	0.94	0.93	0.92	0.92	0.92	0.92	1	0.77	0.903
	Recall	0.89	0.84	0.84	0.83	0.57	0.61	0.57	0.52	0.5	0.93	0.762
	F1	0.83	0.89	**0.89**	0.88	0.7	0.73	0.7	0.66	0.67	0.84	0.827
Frame difference	Precision	0.64	0.64	0.6	0.71	0.65	0.93	0.71	0.71	0.59	0.58	0.676
	Recall	0.95	0.95	0.95	0.94	0.95	0.78	0.95	0.95	0.88	0.93	0.923
	F1	0.76	0.76	0.74	0.81	0.77	0.85	0.81	0.81	0.71	0.71	0.78
Image match	Precision	0.82	0.82	0.72	0.65	0.67	0.63	0.63	0.71	0.55	0.58	0.678
	Recall	0.95	0.95	0.95	0.94	0.95	0.94	0.95	0.95	0.94	0.93	0.945
	F1	**0.88**	**0.88**	0.82	0.77	0.79	0.75	0.76	0.81	0.69	0.71	0.789
Scan line marking	Precision	0.92	0.93	0.93	1	0.94	0.93	0.95	0.93	0.93	0.92	0.938
	Recall	0.63	0.74	0.68	0.61	0.71	0.78	0.86	0.67	0.78	0.8	0.726
	F1	0.75	0.82	0.79	0.76	0.81	0.85	0.9	0.78	0.85	0.86	0.82
SD	Precision	0.94	0.85	0.82	0.74	0.87	0.74	0.91	0.87	0.85	0.82	0.841
	Recall	0.79	0.94	0.94	0.94	0.95	0.94	0.95	0.95	0.94	0.93	0.893
	F1	0.86	0.89	0.88	0.83	0.91	0.83	0.93	0.91	0.89	0.87	0.88
LK	Precision	0.94	0.82	0.78	0.77	0.87	0.74	0.95	0.87	0.85	0.78	0.837
	Recall	0.79	0.94	0.95	0.94	0.95	0.94	0.95	0.95	0.94	0.93	0.928
	F1	0.86	0.88	0.86	0.85	0.91	0.83	**0.95**	0.91	0.89	0.85	0.88
Proposed method	Precision	1	0.85	0.82	0.86	0.91	0.81	0.9	0.91	0.95	0.94	0.895
	Recall	0.79	0.95	0.95	1	0.95	0.94	0.9	0.95	1	1	0.943
	F1	**0.88**	**0.9**	0.88	**0.92**	**0.93**	**0.87**	0.9	**0.93**	**0.97**	**0.97**	**0.918**

Table 7. Comparison of number of detected plankton from 10 randomly selected frames.

The Ten Frames:	1	2	3	4	5	6	7	8	9	10
Top-Hat	19	34	1	47	45	3	3	0	0	2
Frame difference	281	260	10	159	143	15	13	8	10	12
Image match	78	129	12	106	120	16	15	11	9	14
Scan line marking	8	119	1	51	99	7	4	2	2	3
SD	172	195	0	83	124	1	5	1	1	9
LK	163	190	0	86	121	1	5	1	1	10
Proposed method	94	105	0	71	89	1	7	1	1	5
Ground-Truth	87	94	1	66	76	2	6	1	1	5

$$|u + u'| < \beta_1 \quad or \quad |v + v'| < \beta_1 \qquad (24)$$

Secondly, when there is no plankton and the optical flow happens to be small, if the values of the optical flow are not the opposite but the sum still conforms to Equation (24), the threshold β_2 needs to be set to solve this situation, as shown in Equation (25).

$$-uu' > \beta_2 \quad or \quad -vv' > \beta_2 \qquad (25)$$

The best threshold value is obtained by traversing the range value, the scope of β_1 is 0.05 to 0.35, step size is 0.05, the scope of β_2 is 3–9, and the step length is 1. Then, the original images and all those resulting from different thresholds are represented by vectors. At last, we calculate the cosine similarity between two images, that is the calculation of cosine distance between two vectors; the larger the cosine distance between the two vectors, the more similar the two images are. The results are shown in Table 9.

Table 8. Comparison of recall, precision, and F1-score of detected plankton from 10 randomly selected frames.

	The Ten Frames:	1	2	3	4	5	6	7	8	9	10	Average
Top-Hat	Precision	0.95	0.88	1.00	0.94	0.89	0.67	1.00	0.00	0.00	1.00	0.733
	Recall	0.21	0.32	1.00	0.66	0.53	1.00	0.50	0.00	0.00	0.40	0.462
	F1	0.34	0.47	1.00	0.78	0.66	0.80	0.67	0.00	0.00	0.57	0.529
Frame difference	Precision	0.28	0.35	0.1	0.38	0.49	0.13	0.38	0.13	0.10	0.33	0.267
	Recall	0.92	0.96	1.00	0.91	0.92	1.00	0.83	1.00	1.00	0.80	0.934
	F1	0.43	0.51	0.18	0.54	0.64	0.23	0.52	0.23	0.18	0.47	0.393
Image match	Precision	0.90	0.70	0.08	0.57	0.58	0.13	0.33	0.09	0.11	0.29	0.378
	Recall	0.80	0.96	1.00	0.91	0.92	1.00	0.83	1.00	1.00	0.80	0.922
	F1	0.85	0.81	0.15	0.70	0.71	0.23	0.47	0.17	0.20	0.43	0.472
Scan line marking	Precision	0.94	0.73	1.00	0.88	0.71	0.29	1.00	0.50	0.50	0.67	0.722
	Recall	0.86	0.93	1.00	0.68	0.92	1.00	0.67	1.00	1.00	0.40	0.846
	F1	0.90	0.82	1.00	0.77	0.80	0.45	0.80	0.67	0.67	0.50	0.738
SD	Precision	0.47	0.46	0.00	0.72	0.57	1.00	1.00	1.00	1.00	0.56	0.678
	Recall	0.93	0.95	0.00	0.91	0.93	0.50	0.83	1.00	1.00	1.00	0.805
	F1	0.62	0.62	0.00	0.80	0.71	0.67	0.91	1.00	1.00	0.72	0.705
LK	Precision	0.50	0.47	0.00	0.72	0.57	1.00	1.00	1.00	1.00	0.50	0.676
	Recall	0.94	0.95	0.00	0.94	0.91	0.50	0.83	1.00	1.00	1.00	0.807
	F1	0.65	0.63	0.00	0.82	0.70	0.67	0.91	1.00	1.00	0.67	0.705
Proposed method	Precision	0.88	0.86	0.00	0.89	0.81	1.00	0.86	1.00	1.00	1.00	0.830
	Recall	0.95	0.96	0.00	0.95	0.95	0.50	1.00	1.00	1.00	1.00	0.831
	F1	0.91	0.91	0.00	0.92	0.87	0.67	0.92	1.00	1.00	1.00	0.820

Table 9. Select best threshold by comparing cosine distance between two vectors.

Threshold:	0.05	0.10	0.15	0.20	0.25	0.30	0.35
3	0.050	0.059	0.065	0.069	0.072	0.074	**0.076**
4	0.048	0.057	0.062	0.066	0.069	0.072	0.074
5	0.047	0.055	0.061	0.065	0.068	0.070	0.072
6	0.046	0.054	0.059	**0.076**	0.069	0.066	0.070
7	0.045	0.053	0.058	0.062	0.065	0.067	0.069
8	0.044	0.052	0.057	0.061	0.064	0.066	0.068
9	0.044	0.051	0.056	0.060	0.063	0.065	0.067

As shown in Figure 8, Figure 8a is the original image, Figure 8c represents the result of using the threshold β_2, and the one without the threshold β_2 is shown in Figure 8b.

3.4. Time Complexity Comparison

The time complexity comparison of the proposed method and six state-of-the-art methods is provided in Table 10. We select a one-minute video of 1440 frames and calculate the computation time to measure the time complexity of difference methods. Although the proposed method doesn't have great advantage in term of the time complexity, it outperforms other methods in accurate detection of plankton. In terms of the detection efficiency, some experimental comparisons were carried out. Based on the same one-minute video, the computation time and recall rate of the following four different strategies are compared, respectively. We sample pixels at intervals of 1, and take interval frames from full sequence at intervals of 1 frame. According to the results shown in Table 11, the interval between pixels has a weak influence on the error of the result, where the recall rate, precision rate, and *F1-score* are the closest to the original image's result, and the detection efficiency is improved by greatly reducing the calculation time.

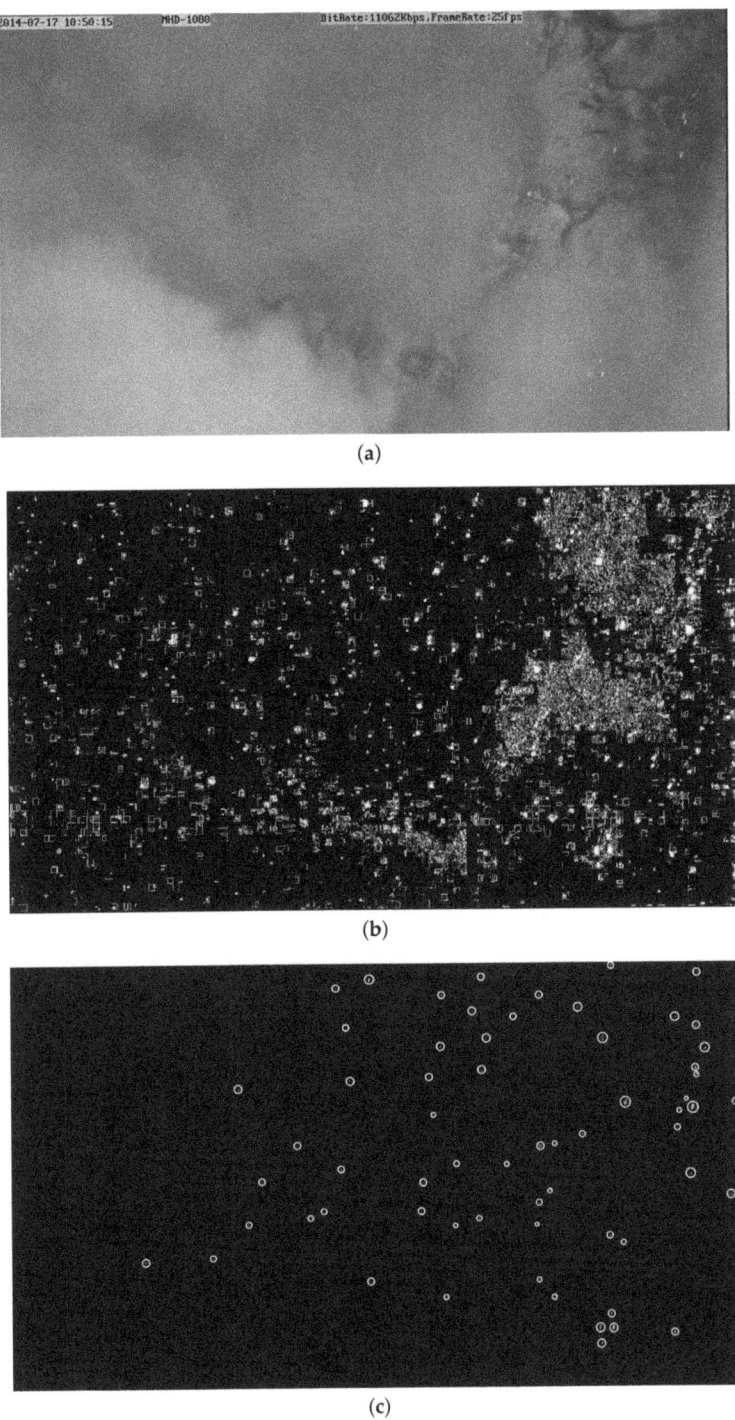

Figure 8. Comparison with or without threshold: (**a**) original image; (**b**) one without threshold β_2, and (**c**) result of using threshold β_2.

Table 10. Time complexity comparison of proposed method and 6 state-of-the-art methods in a 1-min video of 1440 frames.

Top-Hat	Frame Difference	Image Match	Scan Line Marking	SD	LK	Proposed Method
3478 s	176 s	13,149 s	4476 s	989 s	1070 s	1112 s

Table 11. Time complexity comparison of different sampling in a 1-min video of 1440 frames.

A Total of 116 Plankton	Take Interval Frames from full Sequence				
	Quantity	Precision	Recall	F1	Calculation Time
Pixel interval sampling	81	0.86	0.6	0.71	137 s
All the pixels	30	0.83	0.23	0.36	618 s
A Total of 116 Plankton	Full Sequence				
	Quantity	Precision	Recall	F1	Calculation Time
Pixel interval sampling	110	0.95	0.91	0.93	436 s
Full sequence	113	0.97	0.95	0.96	1112 s

4. Conclusions

Detection of plankton plays an important role in the exploration and research of deep-sea areas. Variations in the quantity and spatial distribution of plankton determine the function of the entire marine ecosystem. In this paper, we introduce a method for deep-sea plankton community detection in marine ecosystem with an underwater robotic platform. Compared with that of traditional methods, our method simultaneously improves the precision and recall of plankton detection. The obtained results and the proved theory provide a scientific basis for studying the material cycle and energy flow of deep-sea ecosystems. For our future work, with a view to strengthening the proposed solution, we aim to improve our plankton detection approach, and then conduct studies for plankton recognition and identification of their species.

Author Contributions: Conceptualization, J.W. and Z.D.; methodology, J.W. and M.Y.; writing—original draft preparation, J.W., K.K. and D.W.; visualization, J.R. and K.K.; writing—review and editing J.W., K.K., Q.Z. and J.R. All authors have read and agreed to the published version of the manuscript.

Funding: This research was funded by National Key R&D Program of China under Grant No. 2018YFC0831503.

Institutional Review Board Statement: Not applicable.

Informed Consent Statement: Not applicable.

Data Availability Statement: Not applicable.

Acknowledgments: The authors would like to thank National Deep Sea Center for the provision of the video material used in the experimental validation of the method.

Conflicts of Interest: The authors declare no conflict of interest.

References

1. Brierley, A. Plankton. *Curr. Biol.* **2017**, *27*, 478–483. [CrossRef]
2. Wang, K.; Razzano, M.; Mou, X. Cyanobacterial blooms alter the relative importance of neutral and selective processes in assembling freshwater bacterioplankton community. *Sci. Total Environ.* **2020**, *706*, 135724. [CrossRef]
3. Forster, D.; Qu, Z.; Pitsch, G.; Bruni, E.P.; Kammerlander, B.; Proschold, T.; Sonntag, B.; Posch, T.; Stoeck, T. Lake Ecosystem Robustness and Resilience Inferred from a Climate-Stressed Protistan Plankton Network. *Microorganisms* **2021**, *9*, 549. [CrossRef] [PubMed]
4. Henrik, L.; Torkel, G.; Benni, W. Planton community structure and carbon cycling on the western coast of Greenland during the stratified summer situation. II.Heterotrophic dinoflagellates and ciliates. *Aquat. Microb. Ecol.* **1999**, *16*, 217–232.

5. Lu, H.; Wang, D.; Li, J.; Li, X.; Kim, H.; Serikawa, S.; Humar, I. Conet: A cognitive ocean network. *IEEE Wirel. Commun.* **2019**, *26*, 90–96. [CrossRef]
6. Wiebe, P.; Benfield, M. From the Hensen net toward four-dimensional biological oceanography. *Prog. Oceanogr.* **2003**, *56*, 7–136. [CrossRef]
7. Yan, Y.; Ren, J.; Sun, G.; Zhao, H.; Han, J.; Li, X.; Marshall, S.; Zhan, J. Unsupervised image saliency detection with Gestalt-laws guided optimization and visual attention based refinement. *Pattern Recogn.* **2018**, *79*, 65–78. [CrossRef]
8. Tschannerl, J.; Ren, J.; Yuen, P.; Sun, G.; Zhao, H.; Yang, Z.; Wang, Z.; Marshall, S. MIMR-DGSA: Unsupervised hyperspectral band selection based on information theory and a modified discrete gravitational search algorithm. *Inform. Fusion* **2019**, *51*, 189–200. [CrossRef]
9. Zabalza, J.; Ren, J.; Yang, M.; Zhang, Y.; Wang, J.; Marshall, S.; Han, J. Novel Folded-PCA for improved feature extraction and data reduction with hyperspectral imaging and SAR in remote sensing. *ISPRS J. Photogramm.* **2014**, *93*, 112–122. [CrossRef]
10. Zheng, Q.; Tian, X.; Yang, M.; Wu, Y.; Su, H. PAC-Bayesian framework based drop-path method for 2D discriminative convolutional network pruning. *Multidim. Syst. Sign. Process.* **2020**, *31*, 793–827. [CrossRef]
11. Zheng, Q.; Yang, M.; Tian, X.; Jiang, N.; Wang, D. A Full Stage Data Augmentation Method in Deep Convolutional Neural Network for Natural Image Classification. *Discrete Dyn. Nat. Soc.* **2020**, *2020*, 1–11. [CrossRef]
12. Zheng, Q.; Yang, M.; Yang, J.; Zhang, Q.; Zhang, X. Improvement of Generalization Ability of Deep CNN via Implicit Regularization in Two-Stage Training Process. *IEEE Access* **2018**, *6*, 15844–15869. [CrossRef]
13. Zheng, Q.; Tian, X.; Jiang, N.; Yang, M. Layer-wise learning based stochastic gradient descent method for the optimization of deep convolutional neural network. *J. Intell. Fuzzy Syst.* **2019**, *37*, 5641–5654. [CrossRef]
14. Zheng, Q.; Yang, M.; Zhang, Q.; Zhang, X. Fine-grained image classification based on the combination of artificial features and deep convolutional activation features. In Proceedings of the 2017 IEEE/CIC International Conference on Communications in China, Qingdao, China, 22–24 October 2019; pp. 1–6.
15. Craig, J. Distribution of Deep-Sea Bioluminescence across the Mid-Atlantic Ridge and Mediterranean Sea: Relationships with Surface Productivity, Topography and Hydrography. Ph.D. Thesis, University of Aberdeen, Aberdeen, UK, 2012.
16. Craig, J.; Priede, I.G.; Aguzzi, J.; Company, J.B.; Jamieson, A.J. Abundant bioluminescent sources of low-light intensity in the deep Mediterranean Sea and North Atlantic Ocean. *Mar. Biol.* **2015**, *162*, 1637–1649. [CrossRef]
17. Phillips, B.; Gruber, D.; Vasan, G.; Pieribone, V.; Sparks, J.; Roman, C. First Evidence of Bioluminescence on a "Black Smoker" Hydrothermal Chimney. *Oceanography* **2016**, *29*, 10–12. [CrossRef]
18. Kydd, J.; Rajakaruna, H.; Briski, E.; Bailey, S. Examination of a high resolution laser optical plankton counter and FlowCAM for measuring plankton concentration and size. *J. Sea Res.* **2018**, *133*, 2–10. [CrossRef]
19. Kocak, D.M.; Da Vitoria Lobo, N.; Widder, E.A. Computer vision techniques for quantifying, tracking, and identifying bioluminescent plankton. *IEEE J. Oceanic Eng.* **2002**, *24*, 81–95. [CrossRef]
20. Mazzei, L.; Marini, S.; Craig, J.; Aguzzi, J.; Fanelli, E.; Priede, I.G. Automated Video Imaging System for Counting Deep-Sea Bioluminescence Organisms Events. In Proceedings of the 2014 ICPR Workshop on Computer Vision for Analysis of Underwater Imagery, Stockholm, Sweden, 24 August 2014; pp. 57–64.
21. Schmid, M.; Cowen, R.; Robinson, K.; Luo, J.; Briseno-Avena, C.; Sponaugle, S. Prey and predator overlap at the edge of a mesoscale eddy: Fine-scale, in-situ distributions to inform our understanding of oceanographic processes. *Sci. Rep.* **2020**, *10*, 1–16. [CrossRef] [PubMed]
22. Horn, B.K.P.; Schunck, B.G. Determining optical flow. *Artif. Intell.* **1981**, *17*, 185–203. [CrossRef]
23. Iversen, M.H.; Nowald, N.; Ploug, H.; Jackson, G.A.; Fischer, G. High resolution profiles of vertical particulate organic matter export off Cape Blanc, Mauritania: Degradation processes and ballasting effects. *Deep Sea Res. Part I Oceanogr. Res. Pap.* **2010**, *57*, 771–784. [CrossRef]
24. He, F.; Hu, Y.; Wang, J. Texture Detection of Aluminum Foil Based on Top-Hat Transformation and Connected Region Segmentation. *Adv. Mater. Sci. Eng.* **2020**, *3*, 1–7. [CrossRef]
25. Wei, H.; Peng, Q. A block-wise frame difference method for real-time video motion detection. *Int. J. Adv. Robot. Syst.* **2018**, *15*, 1–13. [CrossRef]
26. Yang, S.; Shi, X.; Zhang, G.; Lv, C.; Yang, X. Measurement of 3-DOF Planar Motion of the Object Based on Image Segmentation and Matching. *Nanomanuf. Metrol.* **2019**, *2*, 124–129. [CrossRef]
27. Wang, W.; Wang, W.; Yan, Y. A Scan-Line-Based Hole Filling Algorithm for Vehicle Recognition. *Adv. Mater. Res.* **2011**, *179*, 92–96. [CrossRef]
28. Lu, N.; Wang, J.; Wu, Q. An optical flow and inter-frame block-based histogram correlation method for moving object detection. *Int. J. Model. Identif. Control* **2010**, *10*, 87–93. [CrossRef]
29. Mohiuddin, K.; Alam, M.; Das, A.; Munna, T.; Allayear, S.; Ali, H. Haar Cascade Classifier and Lucas-Kanade Optical Flow Based Realtime Object Tracker with Custom Masking Technique. *Adv. Inf. Commun. Netw.* **2019**, *2*, 398–410.

Article

A Robotic Experimental Setup with a Stewart Platform to Emulate Underwater Vehicle-Manipulator Systems

Kamil Cetin [1,2,*,†], Harun Tugal [1,2,‡], Yvan Petillot [1,2], Matthew Dunnigan [1,2], Leonard Newbrook [1,2] and Mustafa Suphi Erden [1,2]

[1] Institute of Sensors, Signals and Systems, School of Engineering and Physical Sciences, Heriot-Watt University, Edinburgh EH14 4AL, UK
[2] Edinburgh Centre for Robotics, Edinburgh EH14 4AL, UK
* Correspondence: kamil.cetin@ikcu.edu.tr
[†] Current address: Department of Electrical and Electronics Engineering, Izmir Katip Celebi University, Izmir 35620, Turkey.
[‡] Current address: RACE/United Kingdom Atomic Energy Authority (UKAEA) Culham Science Centre, Abingdon OX14 3DB, UK.

Abstract: This study presents an experimental robotic setup with a Stewart platform and a robot manipulator to emulate an underwater vehicle–manipulator system (UVMS). This hardware-based emulator setup consists of a KUKA IIWA14 robotic manipulator mounted on a parallel manipulator, known as Stewart Platform, and a force/torque sensor attached to the end-effector of the robotic arm interacting with a pipe. In this setup, we use realistic underwater vehicle movements either communicated to a system in real-time through 4G routers or recorded in advance in a water tank environment. In addition, we simulate both the water current impact on vehicle movement and dynamic coupling effects between the vehicle and manipulator in a Gazebo-based software simulator and transfer these to the physical robotic experimental setup. Such a complete setup is useful to study the control techniques to be applied on the underwater robotic systems in a dry lab environment and allows us to carry out fast and numerous experiments, circumventing the difficulties with performing similar experiments and data collection with actual underwater vehicles in water tanks. Exemplary controller development studies are carried out for contact management of the UVMS using the experimental setup.

Keywords: underwater vehicle–manipulator system; robotics emulator; contact management; remote inspection; force control

1. Introduction

An underwater vehicle–manipulator system (UVMS) consists of an underwater robotic manipulator mounted on an underwater vehicle typically used for subsea inspection and surveillance [1–3]. Due to the inherent danger of manned subsea operations, the research interest in underwater robotic systems has continuously increased as UVMSs have a wide range of application areas—for instance, for object inspection, underwater welding, and valve manipulation within the offshore industry [1,2]. The underwater robot manipulators enhance capabilities of the underwater vehicles and reduce operational costs and danger to human life for the essential subsea tasks requiring physical interaction. However, designing robust controller for such a complex system is a challenge from a control point of view due to the highly dynamic coupling between the manipulator and the floating vehicle. In addition, the overall system needs to be robust against external disturbances, e.g., caused by waves or tidal streams, while the end-effector of the manipulator is interacting with the environment. These are also common problems for land-based mobile manipulators but are particularly relevant to UVMSs as the base vehicle is floating. For testing and demonstration purposes, here we consider underwater asset inspection/manipulation tasks which require maintaining a physical contact with the asset surface, where the exact location of the contact on the surface or the exact trajectory followed on the surface is not critical. This is typically the case with pipe thickness measurements, corrosion measurements, cleaning of surfaces from biological structures, and placement of (e.g., magnetically attached) sensors/devices on the surface.

In this study, it is assumed that vehicle motions, caused by environmental disturbances, are unknown for the robotic arm controller while keeping the end-effector in contact with a surface under disturbances. We have emulated environmental disturbances with realistic data that we have collected from a physical underwater vehicle floating in a water tank under the occasional impact of push movements. In addition, the physical interaction of the manipulator end-effector with the surface acts as disturbances in the motion of the vehicle. This is due to the physical coupling between the manipulator and the vehicle and transmission of the interaction force to the base through robot links. The position disturbance due to this force-impact has been computed and applied on the emulating base platform. In this way, we have obtained a realistic emulation of the underwater disturbance impacts on the robot base, by capturing the two main causes: water flow and environment interaction. As a result, a simulation environment and a physical experimental setup have been developed to interact with each other to replicate a UVMS in order to test and validate the controllers in a dry-lab environment. A force/position control method is adapted from our earlier studies [4] and an admittance based controller [5] that applies virtual dynamics at the manipulator end-effector for perpendicular force interaction with the unknown surface is implemented in this study. This admittance controller does not require knowledge of the vehicle position/velocity, the stiffness of the environment or manipulator base disturbance effects. We demonstrate the use of the setup to replicate costly underwater experiments, through an evaluation of an admittance-based controller in comparison to a PID based controller, both in simulations and in physical experiments with the hardware-based emulator.

For the problem of physical contact and surface tracing using a UVMS in the underwater environment, the authors in [6] proposed an optimized redundancy resolution scheme for operational space tracking control of the end-effector of a UVMS. In [7,8], the authors used task-priority-based redundancy resolution methods where the primary task was defined by the operational space position/velocity tracking and force tracking was proposed as a secondary objective. In [9,10], the authors proposed force/position hybrid controllers for the interaction of the end-effector of UVMS with an underwater environment. In [11], an impedance control focused on task priority redundancy solution was developed for contact force control of UVMS. However, these approaches do not consider the problems related to the disturbance effects on the underwater vehicle motion, since they always have access to the position data of the end-effector relative to an inertial base.

In [12–14], the problem of the physical interaction has been considered for the aerial robots, and they developed variable impedance controllers based on force estimations without using force sensors. For the general problem of hard contact interaction of robot manipulators, the authors in [15–17] developed dynamic adaptive hybrid impedance controllers.

In the surveys of underwater robotics [18,19], there are several simulators for the development of underwater robotics. In the TRIDENT project [20], an ROS-based open-source kinematic simulator, named UWSim, was developed for underwater robotics simulation. In [21], Gazebo was integrated into the UWSim to simulate kinematics and dynamics of underwater robots. In [22], the authors extended a Gazebo-based Unmanned Underwater Vehicle (UUV) simulator by implementing the model of hydrodynamic effects. In [10], the authors developed a hybrid simulator for underwater vehicles and manipulators with the ability to accurately simulate hydrodynamic and contact forces of the UVMS with the environment. However, these studies focused only on the development of software-based simulation frameworks to simulate the dynamics or kinematics of underwater vehicles and manipulators. However, due to the complexity of accurately modeling and simulating the physical disturbances and the interaction forces/torques with an environment, a hardware-based emulation system with physical interaction provides more realistic means of testing and validation for a UVMS. Therefore, in our study, in addition to the software-based simulation, we have a hardware emulation of underwater robotics.

Briefly, we can summarize the main contributions of this study as follows: first, we used realistic underwater vehicle movements transmitted in real time in the experimental setup or pre-recorded in a water tank environment. Next, we simulated the water current effect on floating base vehicle motion, considering both hydrodynamic and contact interaction effects. We also used a physical robotic setup with a Stewart platform and a robotic arm manipulator to emulate a UVMS. We then demonstrate the use of this system to perform fast and numerous experiments to compare control schemes for underwater asset inspection without lengthy and costly underwater experiments.

2. Realistic Real-Time Data Set and Transfer from Water Tank to the Land Robotic Setup

In this study, a real Falcon underwater ROV is deployed at sea in a realistic environment. This vehicle is connected through 4G to the remote lab (approximately 160 km apart) where its position and velocity (in 6 DOF) are used to drive a 6 DOF Stewart platform, see Figure 1. This setup provides a good proxy for the real experiments without the need for complex and expensive underwater hardware and integration.

As shown in Figure 1, the land robotic setup was located in the laboratory (in Edinburgh, UK) and real-time communication between the laboratory and the remote water tank (in Blyth, UK) was established through 4G routers (DrayTek Vigor 2862). During the exemplary studies, a time delay of about 0.3 s was observed. The ROV navigation data were recorded during the experiments and are reproducible on the robotic setup to evaluate future algorithm improvements.

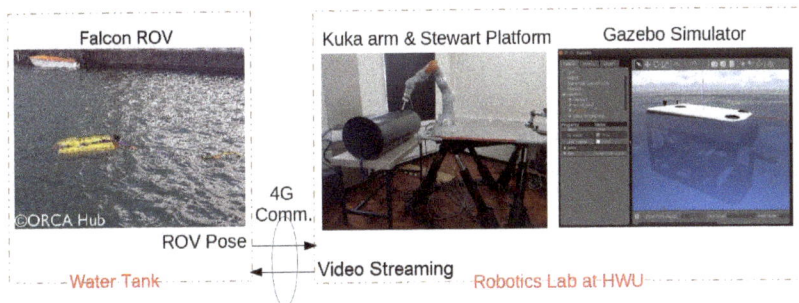

Figure 1. Software and hardware implementations from a real demonstration between the Robotics laboratory in Edinburgh, UK and the water tank in Blyth, UK.

3. Software-Based Simulation Platform

We have developed a UVMS simulation platform in Gazebo using an underwater vehicle and environment proposed in [22]. The simulation platform consists of a 7 DOF robot manipulator model (KUKA IIWA14) mounted on a 4 DOF underwater vehicle model (Rexrov2) and a pipe as an interaction object in the underwater environment. The force sensor attached at the end-effector of the manipulator allows us to measure the interaction force which is used to generate joint motion commands during the surface inspection. In order to move the Rexrov2 in the simulation, Gazebo uses the actual position measurements of the real Falcon ROV in the water tank. Figure 2 shows the overall underwater simulation platform; this platform was developed in Gazebo simulating a UVMS (a robot manipulator mounted on the Rexrov2 vehicle) to perform a surface inspection on a pipe. This simulation platform has been used in integration with the physical setup during the exemplary studies for controller development of contact management.

The simulator we developed is based on the UUV Simulator [22,23] consisting of Gazebo/ROS plugins with the implementation of Fossen's equations of motion for underwater vehicles [24], 6 DOF PID controllers for ROV thrusters' modules, ocean wave model with hydrodynamics and hydrostatic effects, and the Rexrov2 vehicle model [25]. In this way, our physical land robotic setup that will be explained next considers the impact of (simulated) water dynamics and manipulator force interaction effects on the base vehicle, along with other pre-recorded realistic position disturbances.

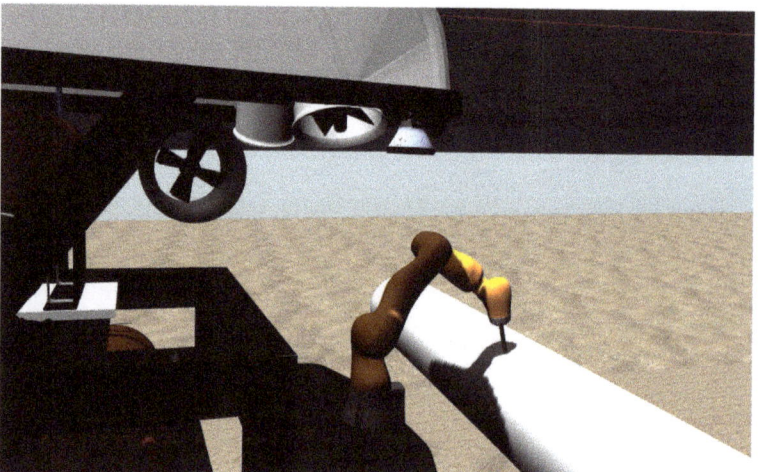

Figure 2. Simulating the UVMS using robotics simulation platform Gazebo. A KUKA IIWA manipulator model mounted on a Rexrov2 vehicle carries out surface inspection on a pipe.

4. Physical Robotic Setup

Figure 3 shows the land robotic setup; this setup emulates a UVMS with a real KUKA IIWA14 robot manipulator fixed on the Stewart parallel manipulator platform interacting with a pipe. It is composed of a 7 DOF robot manipulator (KUKA IIWA14) to emulate an underwater robotic manipulator, a 6 DOF base vehicle (Stewart parallel manipulator) to emulate an underwater vehicle and an ATI Gamma NET FT force sensor attached to the end-effector of the manipulator for the contact management. Since pipes are one of the most common objects to be interacted within the offshore subsea environment [26,27], a PVC vent pipe with a diameter of 500 mm and a thickness of 4 mm was placed in front of the land robotic setup to emulate the underwater object that the UVMS's end-effector is supposed to inspect. In the exemplary studies, the real Falcon ROV's actual position data from the water tank was used to move the Stewart platform. It should be noted that the actual position measurement of the Falcon ROV was only used to move the platform and not to control the manipulator. Since the communication is unilateral and open-loop control is implemented on the Stewart platform, the communication time delay between the two locations did not impact the test and verification of control quality.

Figure 3. Simulating the UVMS with a real KUKA IIWA14 robot manipulator fixed on the Stewart parallel manipulator platform interacting with a pipe.

5. Interaction of Physical Robotic Setup-Realistic Data-Simulation Platform

Generally in a UVMS, once the end-effector of the manipulator contacts an object, the interaction forces and torques at the contact point would result in reaction forces (and torques) on a floating base vehicle that disturbs its position (and orientation) with respect to the inspected object. Therefore, in our physical robotic setup, the interaction forces at the end-effector of the (KUKA) manipulator should be accounted for and reflected to the (Stewart) base platform as a position disturbance. In the simulation platform, we simulated the position disturbance on the floating vehicle due to the real-time force interaction of the end-effector, using the model of a Rexrov2 vehicle with dynamic parameters and PID controllers on its thrusters [22,25]. Afterwards, we embedded these disturbances on top of the previously recorded water wave disturbances (realistic data set) as shown in Figure 4. While the water wave disturbances were pre-recorded, the disturbances due to interaction were dynamically changing in real-time according to the actual interaction of the manipulator in the physical robotic setup. For that purpose, first the force/torque (F/T) interaction that would occur between the underwater manipulator base and vehicle are computed using the end-effector F/T measurements, and then the resultant F/T on the center of mass of the vehicle are computed and superimposed on the force and torque resulting from the thrusters of the Rexrov2 in the simulator. The overall computed movement of the Rexrov2 in the simulator is added to the recorded realistic movement of Falcon ROV in the water tank, and the result is finally transferred to the physical Stewart platform emulating the vehicle movement in the dry-lab.

Overall, we measure the force at the tool-tip in the physical robotic setup and feed this measurement into the simulation platform. The simulator computes the movement of the base under this impact (the simulator considers the models of the robot arm [28] and the base vehicle [23] along with the water dynamics [22,24]). We then merge the simulator vehicle position with the designed disturbance effect (i: no disturbance, ii: sinusoidal movement in each direction, iii: realistic underwater disturbance recorded on an underwater vehicle; as will be explained in the following sections) and send the merged position signal in a feed-forward way to the Stewart platform in the physical robotic setup.

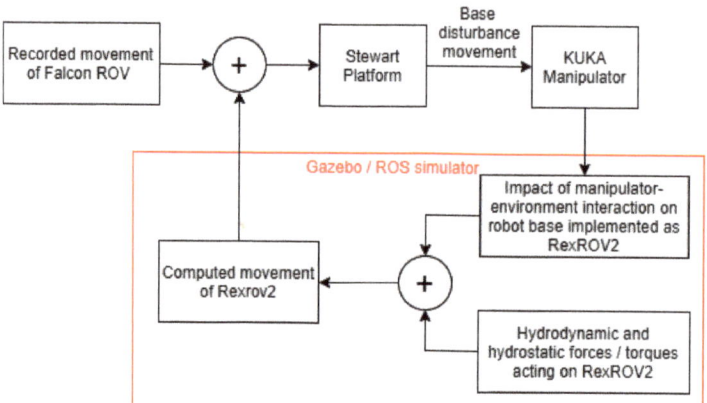

Figure 4. Block diagram of the floating base (Stewart platform) movement.

The closed-loop force/position controllers in the operational space are applied only to the KUKA manipulator for the contact management. On the other hand, the floating base (Stewart platform) is independently controlled by the open-loop position commands provided from real position data of the Falcon ROV due to water wave disturbance and simulated position data of the Rexrov2 due to contact interaction disturbance. All the software implementation of the real-time controllers of the robotic setups, reading of the F/T measurements of the sensor, interacting with the Gazebo simulator, and communicating with the ROV in the water tank through 4G routers was conducted in C++ under Ubuntu with the Robot Operating System (ROS) middleware running at 1 kHz. A marker was attached to the end-effector through a compliant adapter. When the end-effector tool contacts and makes a tracing movement on the pipe surface, the ATI's Gamma F/T sensor attached between the end-effector and the tool measures the forces and torques in 3 translational directions $[x\ y\ z]$ and three rotational directions $[\alpha\ \beta\ \gamma]$ in the operational space at the frequency of 1 kHz. The KUKA robot manipulator uses the KUKA Robot Controller (KRC) that operates at 1 kHz as a client on a remote workstation. The Stewart platform is connected to a real-time QNX control box running at 30 Hz which in turn connects to the central control computer.

6. Exemplary Studies for Development of Contact Management Controllers

The experimental setup was evaluated with the force/position hybrid control architectures of [4,5] for the contact management. The aim of the force controller is to ensure that the end-effector of the robot manipulator is in contact with the environment perpendicularly via applying a linear reference force in the z translational direction (a dynamically changing direction always perpendicular to the unknown surface) and a zero torque in roll (α) and pitch (β) rotational directions in the local (tool) frame. Additionally, the position controller enables the end-effector to follow the desired motion in the x and y directions in the local frame. In these hybrid control methods, the force and position controls are designed independently in dynamically changing local frame directions according to the shape of the surface to generate the end-effector velocity commands in each iteration. This approach is an adaptation of the operational space formulation proposed in [29]. The control strategy in [4] is for fixed-based robot manipulators where a standard proportional (P) controller was used to control perpendicular force interaction and surface trajectory tracking. In [5], taking into account the unknown disturbance effect of the floating base vehicle to the position of the robot manipulator, the control architecture is enhanced via an admittance control approach.

The proposed system has been evaluated in three different application scenarios where in each case the platform commanded to carry out distinct motions (i: no movement on the Stewart platform; *ii*: sinusoidal movement in each Cartesian direction with a position change of $0.1\sin(2\pi Tt)$ m in x, y, z translational and $0.1\sin(2\pi Tt)$ rad in α, β, γ rotational directions with $T = 8$ ms sampling period; *iii*: the actual Cartesian pose of a real ROV submerged in a water tank). In scenarios II and III, Rexrov2's position in the simulator is also added to the movement of the Stewart platform to account for the disturbance effects of hydrodynamics and contact interaction on base vehicle movement. In all scenarios, the performance comparison between the admittance controller [5], the P force controller in [4] and the PID force controller are presented. It should be noted that, when these force controllers are separately implemented in the end-effector's z translational, α and β rotational directions, simultaneously the same PD position controller is implemented to the end-effector's x and y translational directions in all scenarios. For the admittance controller for the perpendicular force contact interaction, the general inertia and damper coefficients were chosen as 0.5 Kg and 100 Ns/m, respectively. For comparison purposes to the case of force control, the PID control gains were used as $K_P = 0.05$, $K_D = 0.5$, and $K_I = 0.002$.

6.1. Application Scenario-I

In the first experiment, the platform is fixed in the global frame for benchmarking. The P, PID, and the admittance controllers are separately implemented on the manipulator for force control. As shown in Figures 5a,b and 6a, the end-effector perfectly tracks the pre-specified trajectory as projected on the 3D surface, and Figure 7c illustrates that it continuously applies the desired force -2 N on the pipe surface.

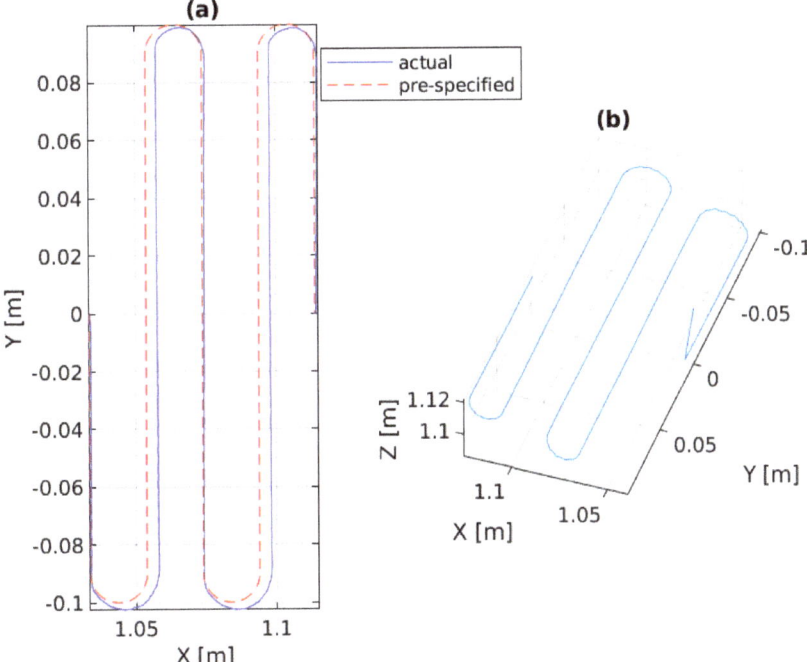

Figure 5. Experimental results of the admittance controller in Scenario-I: (**a**) the 2D pre-specified trajectory on XY plane versus the 2D projection of the 3D trajectory tracking, (**b**) the 3D actual end-effector trajectory on pipe.

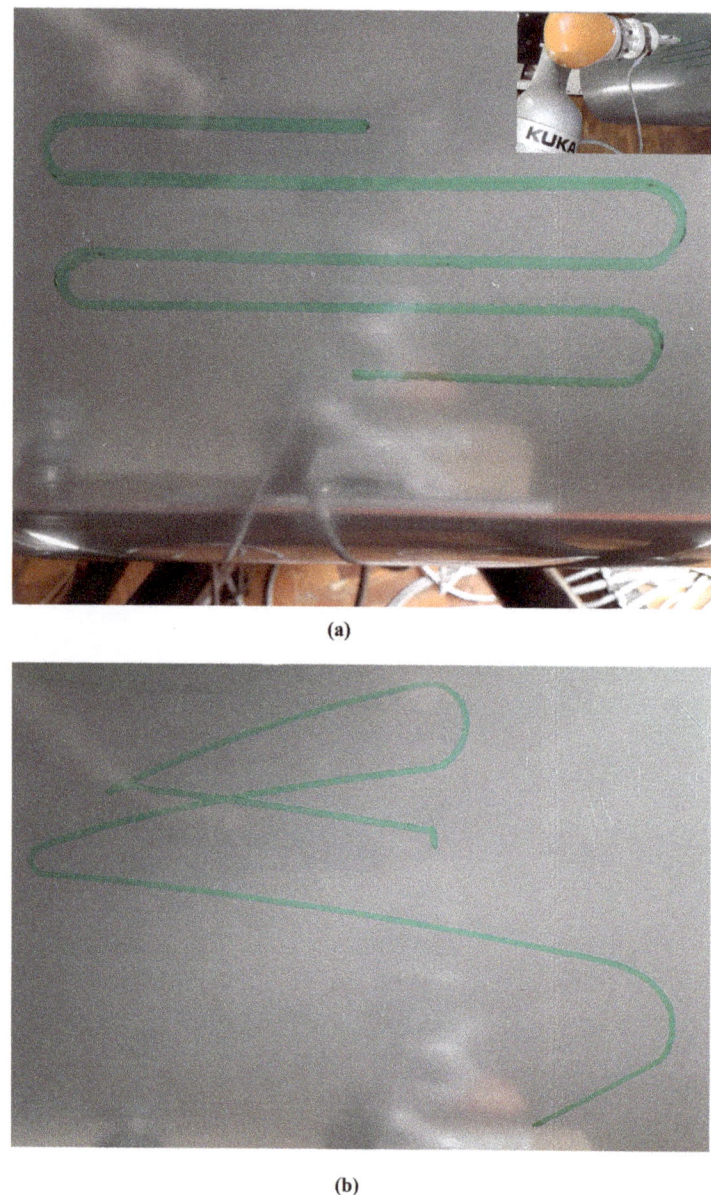

Figure 6. Trajectory drawing pictures on the pipe (the admittance controller was implemented): (**a**) fixed-based manipulator in Scenario-I, (**b**) floating-based manipulator in Scenario-II.

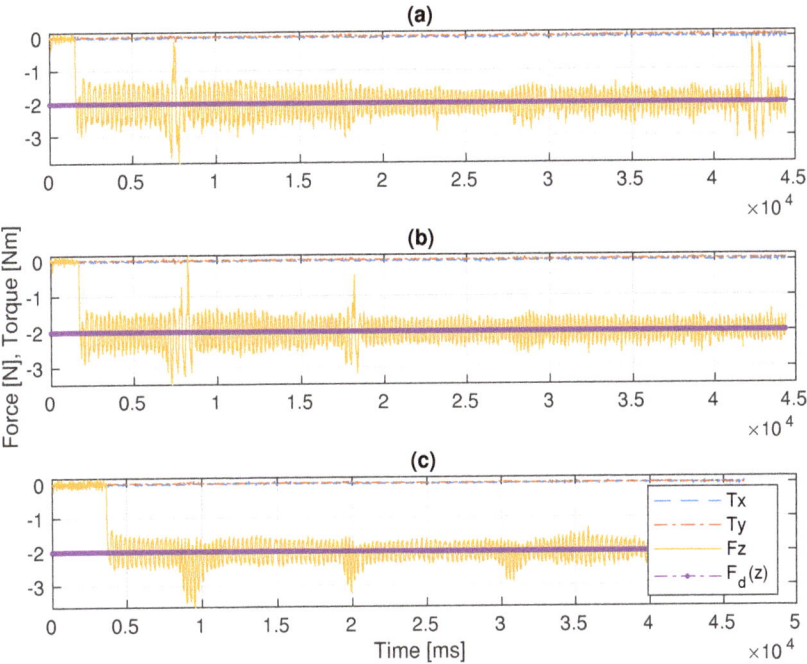

Figure 7. Comparative results of the F/T measurements in Scenario-I: (**a**) the P controller, (**b**) the PID controller, (**c**) the admittance controller.

6.2. Application Scenario-II

In this scenario, a pre-defined sinusoidal Cartesian position along with the Rexrov2 movement due to the hydrodynamic and contact interaction forces effecting the base is commanded to the platform. The purpose here is to observe the manipulator behavior when the vehicle is subject to a known (sinusoidal) disturbance movement (without the complicated disturbance movement of realistic underwater data and without the impact of force interaction of the manipulator). The P, PID, and the admittance controllers are separately implemented on the manipulator for force control. Then, the results of the three force control methods are compared; see Figures 6b and 8. The sinusoidal movement deviates the end-effector trajectory from the intended raster movement. However, as expected, the end-effector remains in contact with the pipe staying perpendicular to the surface and applies a force in the z direction (Figure 9c), no matter how far it deviates from the pre-specified trajectory in the x and y directions (Figures 6b and 8).

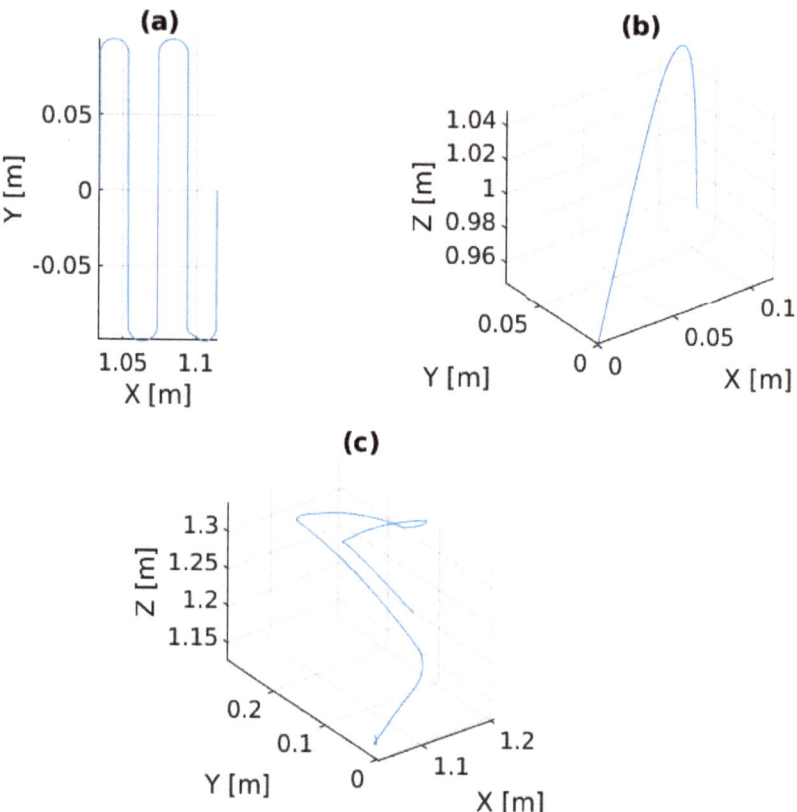

Figure 8. Experimental results of the admittance controller in Scenario-II: (**a**) the 2D pre-specified trajectory on XY plane, (**b**) the 3D vehicle movement as disturbance effects to the robot manipulator, (**c**) the actual end-effector trajectory on pipe with respect to the global frame.

6.3. Application Scenario-III

In this scenario, the Stewart platform moves according to the actual Cartesian pose of the real ROV in the water tank plus the Rexrov2 movement in the Gazebo simulator due to the contact interaction disturbance. Here, as in the previous scenarios, the P, PID, and the admittance controllers are separately implemented to the floating-based manipulator system. Figures 10 and 11c show the 3D actual end-effector trajectories on the pipe for the admittance controller. The movement of the Stewart platform produces a disturbance effect to the base of the KUKA manipulator, but the admittance controller still keeps the end-effector perpendicularly in contact with the pipe as shown in Figure 12c and completes the trajectory tracking within the working space of the pipe surface. However, unlike in the previous scenarios (I and II), the P and PID controllers fail to maintain continuous end-effector contact in the presence of realistic disturbances. Since the Stewart platform mimics the ROV motions through the water wave disturbance and contact interaction disturbance, the actual trajectory tracking positions of the end-effector are different from the pre-specified trajectory.

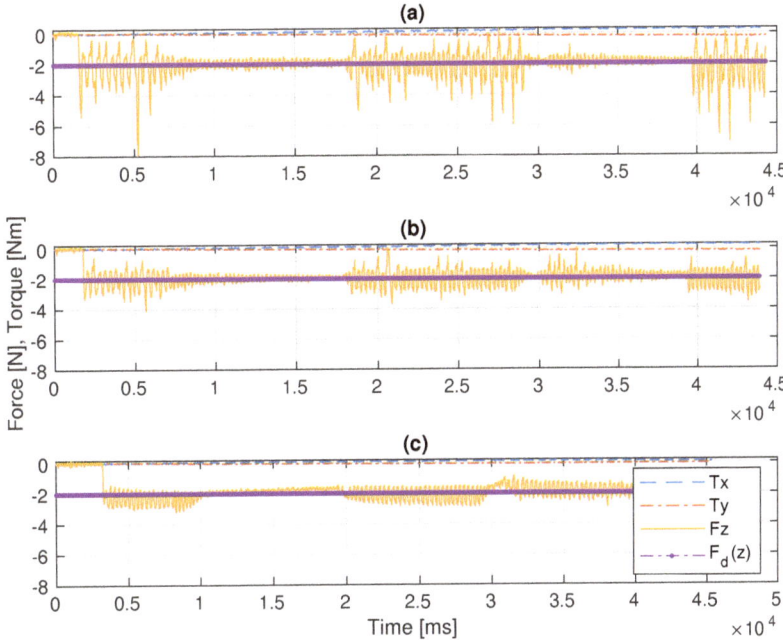

Figure 9. Comparative results of the F/T measurements in Scenario-II: (**a**) the P controller, (**b**) the PID controller, (**c**) the admittance controller.

6.4. Discussion

Before the evaluations, the PID gains were tuned in order to get the best performance possible. The main challenge was to manage the trade-off between stability in contact and fast recovery in case of loss of contact with the surface. For instance, when the system lost contact between the end-effector and the pipe surface, a low P gain resulted in the controller taking significant time to recover the contact. On the other hand, when the robot's end-effector was in contact with the pipe, a large P resulted in instability and frequent cycles of loss-and-recovery of the contact. Therefore, by trial-and-error, the best PID control gains that gave better results than the pure P controller were identified. While the base of the robot manipulator is constantly in motion, the end-effector of the robot manipulator with a highly sensitive force sensor is in constant interaction with an object with an unknown surface and is constantly moving in all directions. Therefore, especially during this interaction, which takes place perpendicular to the surface, the vibrations that occur, as seen in Figures 6, 9 and 11, are caused by the measurements of the very sensitive force sensor. As a result of the advantages of force controllers, these vibrations are minimized.

In Scenario-I, since there is no disturbance on the base movement, the continuous contact and the trajectory tracking of the end-effector is achieved as expected. The mean square force errors $(f(z) - f_d(z))^2$ and the standard deviations in the z perpendicular direction are given in Table 1 for each experimental scenario. The case for Scenario-I constituted a reference in order to compare the impact of disturbances on the base platform. Both the P controller as proposed in [4] and the PID controller designed in this study functioned as well as the admittance controller proposed in [5] (see Figure 7 and Table 1 (I)). However, in Scenario-II, the results with the admittance controller were significantly improved in comparison with the results with the P and PID controllers, (see Figure 9 and Table 1 (II)).

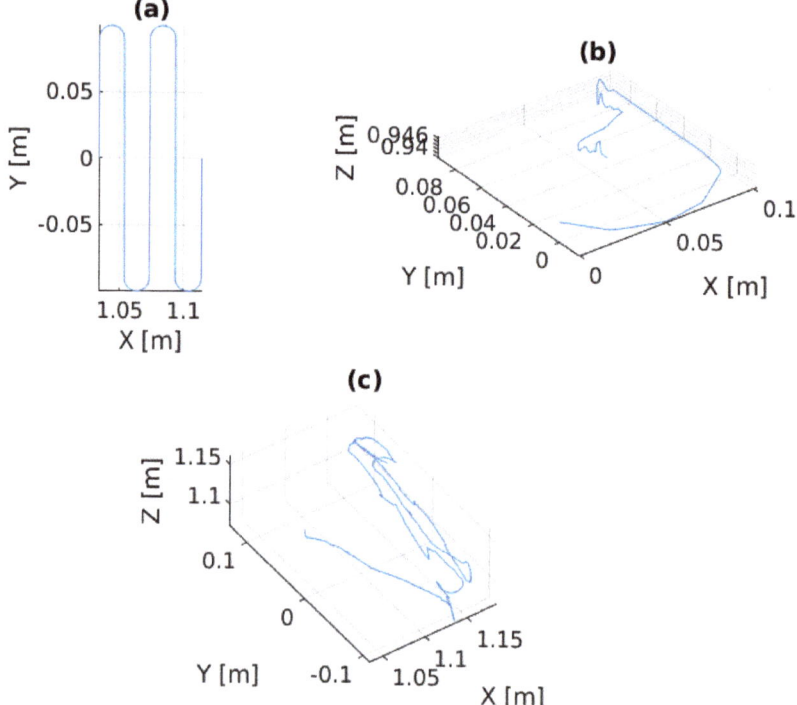

Figure 10. Experimental results of the admittance controller in Scenario-III: (**a**) the 2D pre-specified trajectory on the XY plane, (**b**) the 3D vehicle movement as disturbance effects to the robot manipulator, (**c**) the actual end-effector trajectory on pipe with respect to the global frame.

In the realistic Scenario-III, the controller needs to handle the movements of the base that suddenly change in different directions during the movement of the actual ROV in the water. From Figure 12 and Table 1 (III), it is observed that the deviation from the reference value is significantly less with the admittance controller compared to the other two controllers. In this scenario, various losses of contact with the pipe were observed with all three controllers (see Figure 11). However, the total duration of the loss of contact is much less with the admittance controller (even not observable on the marker trace on the pipe in Figure 11c). It is clearly seen from Figure 11a,b that there are significant losses of contact with the P and PID controllers.

As a result, as seen in Figure 12 and Table 1, when the P, PID, and admittance controllers were compared, the admittance controller has less mean squared force error and standard deviation than the P and PID controllers in fixed-based experimental (I) and floating-based experimental scenarios (II and III). Most importantly, the disturbance effects caused by the floating real ROV and the simulated ROV under the contact interaction can be much better compensated by the admittance controller compared to the P and PID controllers.

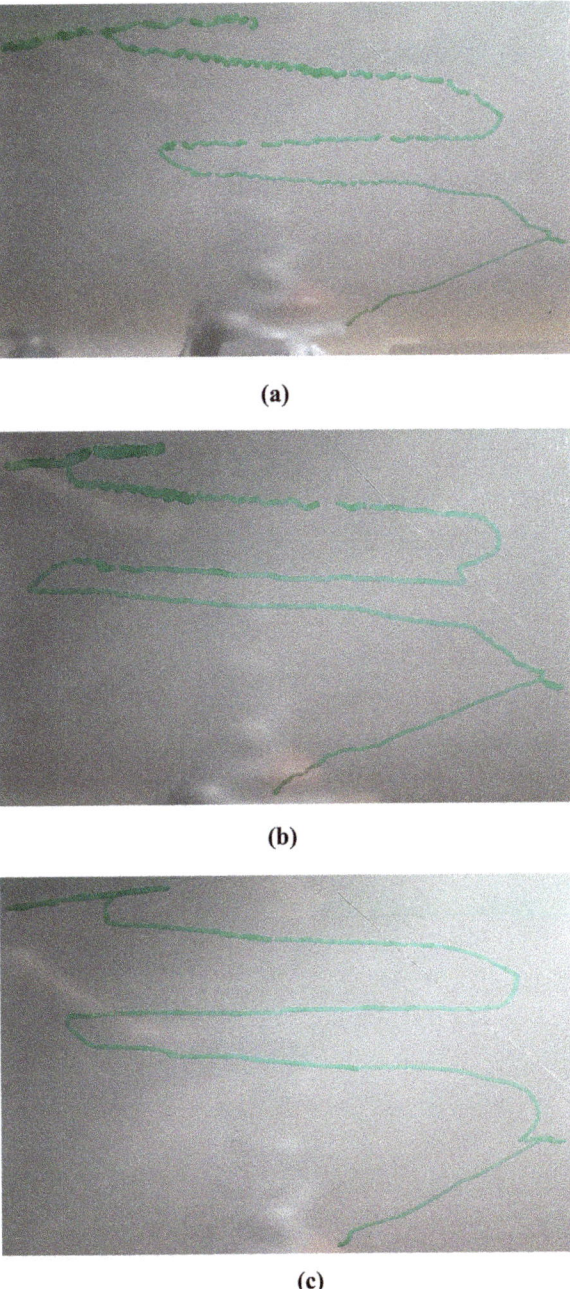

Figure 11. Trajectory drawing pictures on the pipe for floating-based manipulator in Scenario-III: (**a**) implementation for the P controller, (**b**) implementation for the PID controller, (**c**) implementation for the admittance controller.

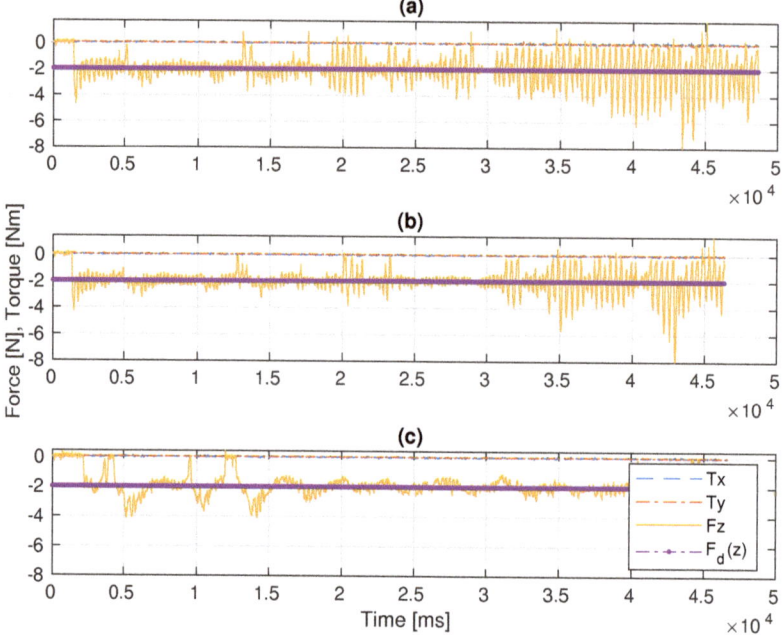

Figure 12. Comparative results of the F/T measurements in Scenario-III: (**a**) the P controller, (**b**) the PID controller, (**c**) the admittance controller.

Table 1. Mean and standard deviations of the squared force errors on the z-direction and the total loss of contact duration from the first contact to the end of the trajectory for the P, PID, and admittance controllers in experimental scenarios.

Application Scenarios	Force Controllers	Mean [N]	Standard Deviation [N]	Loss of Contact Duration [s]
	P	0.19	0.29	-
(I)	PID	0.14	0.22	-
	Admittance	0.11	0.21	-
	P	0.72	2.10	0.695
(II)	PID	0.28	0.41	0.156
	Admittance	0.20	0.23	0
	P	1.13	2.27	4.517
(III)	PID	0.68	2.05	2.274
	Admittance	0.39	0.86	1.783

7. Conclusions

This study demonstrated that force/position control approaches for the physical interaction of the UVMS with underwater structures can be developed with the experimental robotics setup in a dry laboratory environment that allows us to carry out fast and numerous trial experiments. This experimental setup consists of three sub-setups. First, we used realistic underwater vehicle movements transmitted to the system in real time or pre-recorded in a water tank environment. Second, we simulated the water current impact on the floating base vehicle movement considering both hydrodynamic and contact interaction effects. Third, we used a physical robotic setup with a Stewart platform and a robotic arm manipulator to emulate a UVMS. We have demonstrated the use of this system to conduct experiments to compare control schemes for underwater asset inspection, without lengthy and costly underwater experiments. Particularly, we have shown that an admittance control scheme performs better than conventional P and PID controllers for contact and force level management in interaction with an unknown surface.

Author Contributions: Conceptualization, K.C., H.T. and M.S.E.; methodology, K.C., H.T. and M.S.E.; software, K.C. and L.N.; validation, K.C. and H.T.; formal analysis, K.C., H.T. and M.S.E.; investigation, K.C.; resources, Y.P. and M.D.; data curation, K.C.; writing—original draft preparation, K.C.; writing—review and editing, K.C., H.T., Y.P., L.N. and M.S.E.; visualization, K.C.; supervision, M.S.E.; project administration, Y.P. and M.S.E.; funding acquisition, Y.P. and M.S.E. All authors have read and agreed to the published version of the manuscript.

Funding: This work was supported by the UK Engineering and Physical Sciences Research Council (EPSRC) through the ORCA Hub project with Grant No. EP/R026173/1.

Institutional Review Board Statement: Not applicable.

Informed Consent Statement: Not applicable.

Data Availability Statement: Not applicable.

Conflicts of Interest: The authors have no conflict of interest to declare that are relevant to the content of this article. All data, materials and codes are available, and codes were produced by open-source software applications. This manuscript has not been published and is not under consideration for publication elsewhere.

References

1. Dhanak, M.R.; Xiros, N.I. Introduction. In *Handbook of Ocean Engineering*; Springer International Publishing: Berlin/Heidelberg, Germany, 2016; pp. 1–9.
2. Kim, T.W.; Marani, G.; Yuh, J. Underwater Vehicle Manipulators. In *Handbook of Ocean Engineering*; Springer International Publishing: Berlin/Heidelberg, Germany, 2016; pp. 407–422.
3. Sivčev, S.; Coleman, J.; Omerdić, E.; Dooly, G.; Toal, D. Underwater manipulators: A review. *Ocean Eng.* **2018**, *163*, 431–450. [CrossRef]
4. Moura, J.; McColl, W.; Taykaldiranian, G.; Tomiyama, T.; Erden, M.S. Automation of train cab front cleaning with a robot manipulator. *IEEE Robot. Autom. Lett.* **2018**, *3*, 3058–3065. [CrossRef]
5. Cetin, K.; Zapico, C.S.; Tugal, H.; Petillot, Y.; Dunnigan, M.; Erden, M.S. Application of Adaptive and Switching Control for Contact Maintenance of a Robotic Vehicle-Manipulator System for Underwater Asset Inspection. *Front. Robot. AI* **2021**, *8*. [CrossRef] [PubMed]
6. Ismail, Z.H.; Dunnigan, M.W. Redundancy resolution for underwater vehicle–manipulator systems with congruent gravity and buoyancy loading optimization. In Proceedings of the IEEE International Conference on Robotics and Biomimetics, Guilin, China, 19–23 December 2009; IEEE: Piscataway, NJ, USA, 2009; pp. 1393–1399.
7. Olguín-Díaz, E.; Arechavaleta, G.; Jarquín, G.; Parra-Vega, V. A passivity-based model-free force-motion control of underwater vehicle–manipulator systems. *IEEE Trans. Robot.* **2013**, *29*, 1469–1484. [CrossRef]
8. Heshmati-Alamdari, S.; Bechlioulis, C.P.; Karras, G.C.; Nikou, A.; Dimarogonas, D.V.; Kyriakopoulos, K.J. A robust interaction control approach for underwater vehicle manipulator systems. *Annu. Rev. Control* **2018**, *46*, 315–325. [CrossRef]
9. Barbalata, C.; Dunnigan, M.; Petillot, Y. Coupled and Decoupled Force/Motion Controllers for an Underwater Vehicle-Manipulator System. *J. Mar. Sci. Eng.* **2018**, *6*, 96. [CrossRef]
10. Razzanelli, M.; Casini, S.; Innocenti, M.; Pollini, L. Development of a Hybrid Simulator for Underwater Vehicles With Manipulators. *IEEE J. Ocean. Eng.* **2019**, *45*, 1235–1251. [CrossRef]
11. Cieslak, P.; Ridao, P. Adaptive Admittance Control in Task-Priority Framework for Contact Force Control in Autonomous Underwater Floating Manipulation. In Proceedings of the IEEE International Conference on Intelligent Robots and Systems, Madrid, Spain, 1–5 October 2018; IEEE: Piscataway, NJ, USA, 2018; pp. 6646–6651. [CrossRef]
12. Mersha, A.Y.; Stramigioli, S.; Carloni, R. Variable impedance control for aerial interaction. In Proceedings of the IEEE International Conference on Intelligent Robots and Systems, Chicago, IL, USA, 14–18 September 2014; IEEE: Piscataway, NJ, USA, 2014; pp. 3435–3440. [CrossRef]
13. Alexis, K.; Darivianakis, G.; Burri, M.; Siegwart, R. Aerial robotic contact-based inspection: Planning and control. *Auton. Robot.* **2016**, *40*, 631–655. [CrossRef]
14. Ryll, M.; Muscio, G.; Pierri, F.; Cataldi, E.; Antonelli, G.; Caccavale, F.; Bicego, D.; Franchi, A. 6D interaction control with aerial robots: The flying end-effector paradigm. *Int. J. Robot. Res.* **2019**, *38*, 1045–1062. [CrossRef]

15. Duan, J.; Gan, Y.; Chen, M.; Dai, X. Adaptive variable impedance control for dynamic contact force tracking in uncertain environment. *Robot. Auton. Syst.* **2018**, *102*, 54–65. [CrossRef]
16. Cao, H.; Chen, X.; He, Y.; Zhao, X. Dynamic Adaptive Hybrid Impedance Control for Dynamic Contact Force Tracking in Uncertain Environments. *IEEE Access* **2019**, *7*, 83162–83174. [CrossRef]
17. Cao, H.; He, Y.; Chen, X.; Zhao, X. Smooth adaptive hybrid impedance control for robotic contact force tracking in dynamic environments. *Ind. Robot* **2020**, *47*, 231–242. [CrossRef]
18. Matsebe, O.; Kumile, C.; Tlale, N. A Review of Virtual Simulators for Autonomous Underwater Vehicles (AUVs). *Ifac Proc. Vol.* **2008**, *41*, 31–37. [CrossRef]
19. Cook, D.; Vardy, A.; Lewis, R. A survey of AUV and robot simulators for multi-vehicle operations. In Proceedings of the IEEE/OES Autonomous Underwater Vehicles, Oxford, MS, USA, 6–9 October 2014; pp. 1–8. [CrossRef]
20. Prats, M.; Perez, J.; Fernandez, J.J.; Sanz, P.J. An open source tool for simulation and supervision of underwater intervention missions. In Proceedings of the IEEE International Conference on Intelligent Robots and Systems, Vilamoura-Algarve, Portugal, 7–12 October 2012; pp. 2577–2582. [CrossRef]
21. Kermorgant, O. A dynamic simulator for underwater vehicle–manipulators. *Lect. Notes Comput. Sci.* **2014**, *8810*, 25–36.
22. Manhães, M.M.M.; Scherer, S.A.; Voss, M.; Douat, L.R.; Rauschenbach, T. UUV Simulator: A Gazebo-based package for underwater intervention and multi-robot simulation. In Proceedings of the OCEANS 2016 MTS/IEEE Monterey, Monterey, CA, USA, 19–23 September 2016; IEEE: Piscataway, NJ, USA, 2016; pp. 1–8. [CrossRef]
23. Github: UUVSimulator with RexROV2 in Gazebo. Available online: https://github.com/uuvsimulator/rexrov2 (accessed on 2 June 2021).
24. Fossen, T.I. *Handbook of Marine Craft Hydrodynamics and Motion Control*; Willey: Hoboken, NJ, USA, 2011.
25. Berg, V. Development and Commissioning of a DP system for ROV SF 30k. Master's Thesis, Norwegian University of Science and Technology, Trondheim, Norway, 2012.
26. Blyth, W.A. Robotic Pipe Inspection: System Design, Locomotion and Control. Ph.D. Thesis, Imperial College London, London, UK, 2017.
27. Ho, M.; El-Borgi, S.; Patil, D.; Song, G. Inspection and monitoring systems subsea pipelines: A review paper. *Struct. Health Monit.* **2019**, *19*, 606–645. [CrossRef]
28. Github: ROS Kinetic Metapackage for the KUKA LBR IIWA R820. Available online: https://github.com/IFL-CAMP/iiwa_stack (accessed on 2 June 2021).
29. Khatib, O. A Unified Approach for Motion and Force Control of Robot Manipulators: The Operational Space Formulation. *IEEE J. Robot. Autom.* **1987**, *3*, 43–53. [CrossRef]

Article

Fine Alignment of Thermographic Images for Robotic Inspection of Parts with Complex Geometries

Carmelo Mineo [1,*], Nicola Montinaro [2], Mario Fustaino [2], Antonio Pantano [2] and Donatella Cerniglia [2]

[1] Institute for High-Performance Computing and Networking, National Research Council of Italy, 90146 Palermo, Italy
[2] Department of Engineering, University of Palermo, 90128 Palermo, Italy
* Correspondence: carmelo.mineo@icar.cnr.it; Tel.: +39-091-680-9720

Abstract: Increasing the efficiency of the quality control phase in industrial production lines through automation is a rapidly growing trend. In non-destructive testing, active thermography techniques are known for their suitability to allow rapid non-contact and full-field inspections. The robotic manipulation of the thermographic instrumentation enables the possibility of performing inspections of large components with complex geometries by collecting multiple thermographic images from optimal positions. The robotisation of the thermographic inspection is highly desirable to improve assessment speed and repeatability without compromising inspection accuracy. Although integrating a robotic setup for thermographic data capture is not challenging, the application of robotic thermography has not grown significantly to date due to the absence of a suitable approach for merging multiple thermographic images into a single presentation. Indeed, such an approach must guarantee accurate alignment and consistent pixel blending, which is crucial to facilitate defect detection and sizing. In this work, an innovative inspection platform was conceptualised and implemented, consisting of a pulsed thermography setup, a six-axis robotic manipulator and an algorithm for image alignment, correction and blending. The performance of the inspection platform is tested on a convex-shaped specimen with artificial defects, which highlights the potential of the new combined approach. This work bridges a technology gap, making thermographic inspections more deployable in industrial environments. The proposed fine image alignment approach can find applicability beyond thermographic non-destructive testing.

Keywords: robotics; thermography; non-destructive testing; image alignment; image blending

Citation: Mineo, C.; Montinaro, N.; Fustaino, M.; Pantano, A.; Cerniglia, D. Fine Alignment of Thermographic Images for Robotic Inspection of Parts with Complex Geometries. *Sensors* **2022**, *22*, 6267. https://doi.org/10.3390/s22166267

Academic Editor: Ruqiang Yan

Received: 11 July 2022
Accepted: 17 August 2022
Published: 20 August 2022

Publisher's Note: MDPI stays neutral with regard to jurisdictional claims in published maps and institutional affiliations.

Copyright: © 2022 by the authors. Licensee MDPI, Basel, Switzerland. This article is an open access article distributed under the terms and conditions of the Creative Commons Attribution (CC BY) license (https://creativecommons.org/licenses/by/4.0/).

1. Introduction

Non-destructive Testing (NDT) comprises highly multidisciplinary groups of analysis techniques used throughout science and industry to evaluate materials' properties and ensure the integrity of components/structures without causing damage to them [1]. In civil and industrial manufacturing, the increasing deployment of smart/composite materials demands high integrity and traceability of NDT measurements, combined with rapid data throughput. Traditional manual inspection approaches are insufficient in some scenarios since they produce a manufacturing process bottleneck [2]. Therefore, there are fundamental motivations for increasing automation in NDT. Computer-Aided Design (CAD) has been extensively used in engineering design phases. Computer-Aided Manufacturing (CAM) also allows large components to be produced quickly through combinations of traditional subtractive approaches and novel additive manufacturing processes [3]. As a result, large components with complex geometries have become very common in modern structures. NDT inspection is still often performed manually by technicians who typically must move appropriate probes over the contour of the part surface. Manual scanning requires trained technicians and results in a prolonged inspection process for large samples. Automation of NDT is required to cope with the inspection of such structures. Robotic manipulation of

NDT sensors also plays an essential role in inspecting parts made of composite materials. A fundamental issue with composite components is that parts designed as identical can have significant deviations from the CAD model. Composite parts suffer from inherent but different part-to-part springiness out of the mould, which presents a significant challenge for precision NDT measurement deployment. While manual scanning may remain a valid approach for some specific areas of a structure, developing reliable automated solutions has become an industry priority to drive down inspection times and costs. An industrial robot is an automatically controlled, reprogrammable, multipurpose manipulator programmable in three or more axes [4]. Many manufacturers of industrial robots have produced robotic manipulators with excellent positional accuracy and repeatability. In the spectrum of robot manipulators, some modern robots have suitable attributes to develop automated NDT systems. They present precise mechanics, the possibility to accurately master each joint and the ability to export positional data at high update rates. The key challenges to face when developing a robotic NDT system include integrating the NDT instrumentation with the robotic manipulator, creating a suitable robot inspection path for the part under inspection, and developing software for NDT data collection and visualisation. These challenges have been addressed by several applications of six-axis robotic arms for the inspection of parts through automated ultrasonic techniques [5–7]. Robotic ultrasonic inspection has become commonplace thanks to the research investments driven by the aerospace sector in the suitability of ultrasonic techniques to inspect critical aerospace components. Some works have presented robotic ultrasonic inspection systems capable of achieving high data throughputs, accompanied by bespoke software for data visualisation and analysis [5,8]. Automated geometry mapping has also been demonstrated using robotically manipulated metrology sensors [9].

Besides these techniques, other types of inspections have not reached the same level of robotisation; this is the case for thermographic testing, also known as thermal imaging, infrared (IR) thermography or simply thermography. It is an NDT imaging technique that allows the visualisation of heat temporal patterns in an object or a scene and is based on the principle that two dissimilar materials possessing different thermophysical properties produce two distinctive thermal signatures that can be revealed by an infrared sensor, such as an IR thermal camera [10–12]. Although a thermographic setup in reflection mode, with a heat excitation source and an IR camera on the same side of the part under inspection, is not well suited to detect defects located deep in the volume of a component, it presents some advantages over ultrasonic-based inspections. It is contactless and full-field, meaning that the whole area of a component detectable within the field of view of an IR camera is inspected remotely at once. Schmidt and Dutta [13] proposed using industrial robots as manipulators to perform active thermography in 2012. The robotic manipulation of the thermographic instrumentation can enable the possibility of performing inspections of large components with complex geometries by collecting multiple thermographic images at given positions. Despite preliminary investigations [13,14], the robotisation of the thermographic inspection method has not been fully exploited to date due to the lack of a suitable approach capable of aligning automatically-collected thermographic images. The importance of consistent registration of NDT data in CAD models is highlighted in [15]. Aligning thermographic images for NDT analysis is not trivial since accurate and consistent pixel blending must be guaranteed and is crucial to facilitate defect detection and sizing. Inaccurate alignment and blending may create unreal artefacts in the composite thermography image and cause false-positive flaw detection. In this work, an innovative inspection platform was conceptualised and implemented, consisting of a pulsed thermography setup, a six-axis robotic manipulator and a novel algorithm for image transformation, alignment and blending. The performance of the inspection platform is tested on a convex-shaped specimen with artificial defects, highlighting the potential of the new combined approach.

The remaining part of this work is organised as follows. Section 2 reviews the theoretical principles of thermographic inspection and provides scientific literature references. Following a detailed clarification of the origin of the misalignment in robotically-acquired images and the limitations of existing image alignment algorithms, Section 3 describes the novel image alignment and blending algorithm developed by this work. Section 4 introduces the automatic thermography setup used to validate the proposed method. Section 5 presents the experimental results. The outcomes of this work and the method's performance and prospects are discussed in Section 6.

2. Thermography Principles

Thermography, as introduced above, can be deployed through different techniques [16]. The essential equipment for manual (not automated) thermography includes an IR camera, a computer to record (and sometimes process) data and a monitor to display images. The main classification of the thermographic techniques differentiates between passive and active techniques. Passive thermography exploits the fact that materials and structures may naturally be at different (higher or lower) temperatures than the background. For example, the human body is generally at a higher temperature than the ambient; hence it is easily detected by an IR camera without additional stimulation. Conversely, an external stimulus is needed in active thermography to produce a thermal contrast in the object's surface. Active techniques are particularly suited to non-destructive testing since an object containing internal defects (such as voids, delaminations and/or inclusions of foreign material) will require the excitation of thermal disequilibrium to produce a distinctive surface thermal signature detectable with an IR camera. In the realm of active thermographic techniques, pulsed thermography (PT) has broad applicability in NDT. When an object's surface is heated through a short (a few milliseconds) energy pulse of light radiation, a series of thermal waves with different amplitudes and frequencies propagate inside the object medium in a transient mode. The surface temperature is monitored under the principle that defective areas cool down (or heat up) at a different rate than non-defective areas [17–19]. It is known that the thermal wave originating from the energy pulse can be decomposed into a multitude of individual sinusoidal components and that it is possible to link temporal and frequency domains. In pulsed phase thermography (PPT), the PT is combined with the phase and frequency concepts of lock-in thermography (LT), where specimens are subject to a periodical excitation [12,20–22]. Flash lamps generate a heat pulse of high intensity and low duration. The subsequent temperature decay is then acquired over a truncation window.

Once raw data are collected, there are multiple techniques to analyse the data. One approach consists of calculating the Discrete Fourier Transform (DFT) to evaluate the thermal response's frequency content. The phase of specific harmonic content can finally be obtained and presented as a phasegram, an image where the scalar value associated with each pixel represents the phase. Any discontinuity in phase contrast is either caused by the object geometry or indicates a potential flaw. In the PPT approach, whereas deeper anomalies are expected to be better contrasted in low-frequency phasegrams, high-frequency phasegrams probe better for superficial issues. The signal normalisation inherent in evaluating the phase is also expected to reduce the counter effects of non-uniform heat deposition and environmental reflections [23]. It must be noted that the terms phasegram(s) and thermographic image(s) are used as synonyms in the remainder of this paper.

3. Fusion of Multiple Thermographic Images

3.1. Misalignment Issue in Robotically-Acquired Images

Robotic NDT inspections generally occur in a well-structured environment, where the part position is precisely registered with respect to the robot reference system. Great care is dedicated to ensuring the robot tool path is accurately referenced to the sample reference frame to ensure effective data collection during automated inspections [24]. Despite the efforts, a deviation between the actual tool path and the ideal tool path always remains due to the following reasons: (i) the physical tolerances in the robot joints; (ii) the geometric deviations in the mounting support of the sensing instrumentation; (iii) the residual inaccuracy in the calibration of the part position; (iv) the deviation between the actual sample geometry from the part digital counterpart. For these reasons, the resultant data usually reveal some imperfect alignment when they are encoded through robot positional feedback and plotted in the form of a single map. For robotic thermographic inspections, the problem translates to evident misalignment of the thermographic images. The issue may be mitigated through an external metrology tracking system (e.g., a six-DoF laser tracker), capable of measuring the position of the sensing instrumentation with respect to an absolute reference frame. However, such metrology systems are expensive and can increase the overall complexity of robotic inspection systems. In robotic machine vision systems, the exact position of the camera with respect to the robot mounting point is calibrated through the hand-eye calibration method [25], which is based on the knowledge of the camera's intrinsic parameters, such as focal length, aperture, field-of-view and resolution, and on the capture of a calibration pattern (e.g., a checkerboard) from different viewpoints. However, this method is not always applicable to thermographic cameras since they do not usually have a visible-light imaging sensor (RGB sensor). A similar method based on calibration patterns with different thermal infrared emissivity could be adopted to calibrate IR cameras [26]. This work developed a practical solution consisting of correcting each image's plotting location and its prospective aberrations to obtain a misalignment-free full-field view of the inspected sample. The remainder of this section explains the limitations of available image-stitching algorithms and the theoretical foundations of the proposed method herein.

3.2. Limitations of Existing Alignment Methods

Algorithms for aligning images and stitching them into seamless photo-mosaics are among the oldest and most widely used in computer vision. The alignment of images requires establishing mathematical relationships that map pixel coordinates from the unaligned images to their aligned versions. Five parametric 2D planar transformations have been defined [27] (see Figure 1). Each one of these transformations can be described by a transformation matrix ($\tau(p)$), with p being a vector of parameters. Pure translation can be written as $x' = x + t$ or $x' = \tau(p) \cdot x = \begin{bmatrix} I & t \end{bmatrix} \cdot x$, where $x = \{x, y, 1\}$ is the vector of coordinates of the untransformed image pixel and x' denotes the coordinates of the same pixel in the transformed image, I is the (2 × 2) identity matrix, and $t = \begin{bmatrix} t_x & t_y \end{bmatrix}'$ is the translation vector, containing two parameters (respectively, the translation along the x-axis and the translation along the y-axis). The Euclidean transformation is written as $x' = \tau(p) \cdot x = \begin{bmatrix} R & t \end{bmatrix} \cdot x$, where R is the 2D rotation matrix. Thus, Euclidean transformation depends on three parameters: t_x, t_y, and an angle θ (for the rotation matrix). Euclidean distances are preserved. The similarity transformation, also known as scaled rotation, preserves angles between lines. It is expressed as $x' = \begin{bmatrix} sR & t \end{bmatrix} \cdot x$, where s is the scale parameter that brings the parameter counter to four. It must be noted that s is a scalar and the scaling operation is intended to be isotropic. The affine transform is written as $x' = \tau(p) \cdot x = A \cdot x$, where A is an arbitrary 2 × 3 matrix with six parameters. Parallel lines remain parallel under affine transformations. Projective transformation, also known as perspective or homography, is expressed as $x' = \tau(p) \cdot x = H \cdot x$, where H is an arbitrary 3 × 3 matrix:

$$x' = \begin{bmatrix} h_{00} & h_{01} & h_{02} \\ h_{10} & h_{11} & h_{12} \\ h_{20} & h_{21} & 1 \end{bmatrix} \cdot x \qquad (1)$$

Thus, perspective transformation requires eight parameters and preserves straight lines.

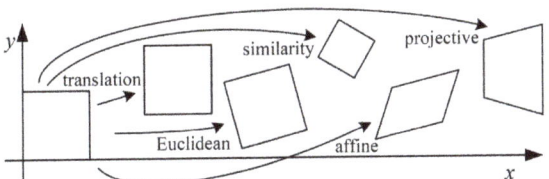

Figure 1. The basic set of 2D planar transformations (Reprinted with permission from Ref. [28]. 2007, now publishers inc).

Assuming the choice of a suitable motion model to transform each image, a typical strategy to align a collection of images consists of aligning the images in pairs. In order to align a pair of images, it is necessary to devise some methods to estimate the parameters to apply the selected transformation to one image while the other is kept fixed. One approach is to shift or warp the first image relative to the other and measure how much the pixels agree. The first methods to quantitatively measure such agreement are often called "direct methods", based on pixel-to-pixel matching [29]. These methods are usually slow since the number of pixel pairs to evaluate can be very large. Direct methods work by directly minimising pixel-to-pixel dissimilarities; a different class of algorithms works by extracting a sparse set of features and then matching these to each other [27,30,31]. Feature-based approaches have the advantage of being more robust against scene movement, are potentially faster, and can be used to automatically discover the adjacency (overlap) relationships among an unordered set of images [32].

Although feature-based approaches work well to create panoramas of scenes with enough distinguishable features, they are not suited to align multiple images for NDT applications. Non-destructive testing aims to detect defects in parts and/or structures. As such, besides the presence of intrinsic geometrical details (e.g., borders and corners), most images may appear relatively featureless since the presence of defects is not the norm. An attempt to use a feature-based alignment approach was presented in [33], where the authors note the need to mark artificial points on the background of a test objective to obtain the mapping matrix from two-dimensional (2D) thermal wave imaging data to the 3D spatial coordinate's digital model. On the other hand, feature-based approaches can also fail if plenty of spatially periodic features are present in the images, which can be the case for industrial components due to stiffeners/stringers, heat dissipators and/or fixturing holes. Direct methods are less prone to failure caused by a lack of image features or abundance of periodicity since they can leverage any consistent low-contrast gradient to find the optimum image transformation parameters. However, the scientific literature does not show any solution readily available to work with the scalar information present in each pixel of thermographic images. As stated above, thermographic images differ from RGB or grayscale images since the pixel values may represent phases (expressed in degrees or radians) and may be negative values. Moreover, the optimum solution to align and stitch multiple thermographic images can not progress pairwise. Although it can work only for images taken in a single row, like in the case of a horizontal panorama, robotic thermographic inspection generally collects images through a raster tool-path, with multiple images arranged in multiple passes. A pairwise image-stitching algorithm would produce a visible drift between adjacent passes due to the progressive summation of alignment errors.

3.3. Fine Pixel-Based Alignment Method

This work developed a direct method capable of simultaneously aligning multiple images. The method is suitable to be used when the rough position of the camera (the shooting pose of each image) is known. That is the case for robotically acquired images, where the camera position is obtained from the robot's positional feedback. Given a set of images, knowledge of camera shooting poses allows skipping the search for the adjacency relationships among the set. Knowing the scale factor makes it possible to convert the pixel index coordinates to real-world coordinates and identify the overlap between the images. The scale factor can easily be calculated by measuring the size of a known object or the known distance between two points in an image in terms of the number of pixels and considering the actual length it represents. Therefore, the algorithm herein is specifically targeted to perform a fine alignment of all images in the set. It is referred to as the Fine Pixel-based Alignment Method (FiPAM). It must be noted that FiPAM is currently suitable for aligning multiple mosaic images of a sample surface that curves only in one direction. Although the constraint of single direction curvature is a significant limitation, it does not impede using FiPAM for mosaic images of any surface belonging to the large family of cylindrical surfaces intended as "generalised cylindrical surfaces" [34]. Under that condition, all collected images can be transposed to a planar domain. Indeed, any cylindrical surface can be represented in the plane by "unrolling" it on a flat surface. An additional assumption is that the part surface captured within the camera field of view is sufficiently close to a flat plane. In other words, the ratio between the local surface curvature and the camera field of view must be small. Figure 2 illustrates a set of nine images used to explain the theoretical foundations of FiPAM.

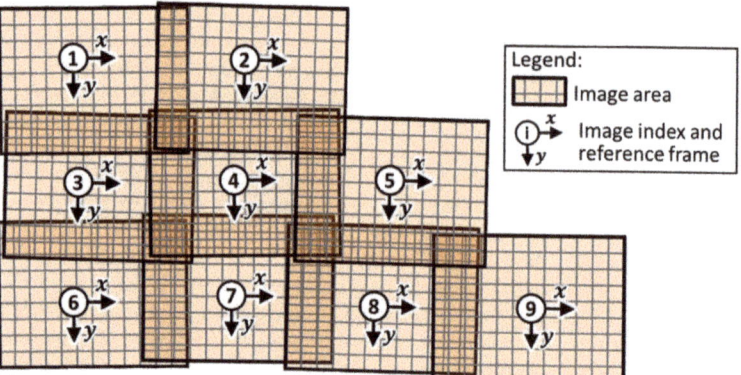

Figure 2. Schematic representation of a set of nine images used to explain the theoretical foundations of FiPAM.

Direct methods find the optimum alignment between a pair of images by an iterative search, where one image is transformed with respect to the other through one of the five planar transformations. To use a direct method, a suitable error metric must first be chosen to measure the goodness of the alignment. Given two images, with one image ($I_0(x)$) taken as a reference image sampled at discrete pixel locations ($x_k = \{x_k, y_k, 1\}$), with k being the pixel index, we wish to find the optimum transformation parameters that align it with the second image ($I_1(x)$), which is kept fixed. The error metric is defined as the sum of squared differences (SSD) of the pixel values of I_1 at the transformed pixel locations and the reference values of I_0. This kind of function has been successfully used in the image processing literature, with different aims (e.g., inpainting [35]). Given a transformation ($\tau(p)$), with p being a vector of parameters, we have:

$$SSD(\tau(p)) = \sum_k [I_1(\tau(p) \cdot x_k) - I_0(x_k)]^2 \tag{2}$$

The optimum set of parameters (p^*) can be found by solving a least-squares problem of this SSD function. Since the transformation allows multiple degrees of freedom (DoFs) for the image, this is a multi-parameter problem. Therefore, a suitable search technique must be devised. The most straightforward technique would be to exhaustively try all possible alignments (full search). In practice, this would be too slow and is not practicable.

Several works have developed hierarchical coarse-to-fine search techniques based on image pyramids [27] when the approximate image alignment is unknown. In this work, since the approximate position of each image is assumed to come from the known camera pose, it has been decided to limit the search space by setting lower and upper bounds for the transformation parameters.

Regarding the set of images in Figures 2 and 3 illustrates all the overlaps between image #4 and its neighbour images. Given a positive scalar value herein named "offset" (o), it is possible to draw shrunk overlap areas whose boundary is at distance o from the boundary of the original overlap areas. The actual value to use for o depends on the expected maximum entity of misalignment caused by the inaccuracy in robotic manipulation of the camera and by the deviations in the physical camera support. Assuming these offset areas move with image #4 and the original overlap areas stick with the parent neighbour image, the bounds of the transformation parameters guarantee that the offset overlap areas remain within the original overlap footprints. Generalising Equation (2) to allow simultaneous alignment of multiple images, FiPAM is based on the following SSD function.

$$SSD(p) = \sum_{i=1}^{n}\sum_{j=1}^{n}\sum_{k}[I_j(\tau_i(p) \cdot x_k) - I_i(\tau_j(p) \cdot x_k)]^2 \quad \text{with } (j \neq i) \text{ and } (k \in K_{i,j}). \quad (3)$$

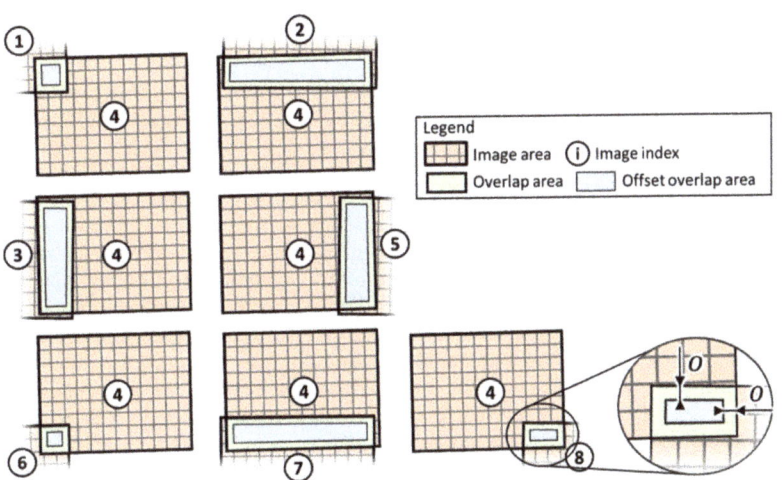

Figure 3. Illustration of all overlap areas between image #4 and neighbour images. The magnified region serves to clarify the relationship between the overlap areas and the offset overlap areas.

$K_{i,j}$ is the set of pixel indices that fall within the offset area, produced by the overlap between the ith and the jth image. Assuming a set of n images, Equation (3) is the sum of squared differences of the pixel intensity values of the jth image and the ith image. Crucially, the overlap pixel locations of the jth image are transformed according to the transformation matrix of the ith image ($\tau_i(p)$) and the locations of the ith image are transformed according to the transformation matrix of the jth image ($\tau_j(p)$).

Now, it must be noted that the vector p includes all the parameters required in the transformation matrices, and only a subset of it is used to compute a single transformation matrix ($\tau_i(p)$, with $i = 1 : n$). Moreover, the summation is not evaluated for $j = i$ (an image is always aligned with itself) and for combinations of i and j corresponding to images that do not overlap, where $K_{i,j}$ is an empty set. This formulation solves a typical problem with pixel-based methods, which is the possibility that parts of I_i may lie outside the boundaries of I_j. This advantage follows directly from the constraints applied to the search space for the transformation parameters. Another aspect to discuss relates to the fact that the transformed pixel indices can be fractional, so a suitable interpolation function must be applied to evaluate the image intensities (I_i and I_j). This work employs bi-cubic interpolation, which yields better results than bilinear interpolants [36]. It must be noted that Equation (3) does not require the image pixel values to be in a specific format. Thus, it can work with the phase values of thermographic phasegrams and images with three RGB colour channels, although it is also possible to first transform the images into a different colour space.

The mathematical parametric formulation of all transformation matrices pictured in Figure 1 was implemented in FiPAM. The formulation allows maximum flexibility in choosing the most suitable transformation for each image, meaning that all images in a set can be aligned using the same type of planar transformation, or each image can use a transformation of a different type. In other words, each image can be transformed by allowing different DoFs, which relate to a different number of parameters. Automating the selection of the optimum transformation for each image is out of the scope of this work. In practical situations, similarity or affine transformations produce satisfactory results if the part surface captured within the camera field of view is sufficiently close to a flat plane. Once the optimum transformation parameters are found, the aligned version of the ith image is computed by transforming its original discrete pixel locations with the following equation:

$$x'_i = \tau_i(p^*) \cdot x_i \tag{4}$$

3.4. Image Blending

Aligning all images in a dataset is not sufficient to merge the images into a single composite image. Indeed, multiple aligned images may present significant differences in pixel intensities in overlapping areas. For RGB images, exposure differences are typically caused by ambient light changes during image capture. In active thermographic imaging, the same problem may be caused by the progressive increase of an object's surface temperature when it is subject to multiple heat pulses. Image blending is usually accomplished through averaging the intensity of homologue/overlapping pixels or by using more sophisticated methods, such as "Laplacian pyramid blending" [37] and "Gradient-domain blending" [38]. Although these blending methods work well and have been implemented in many variants for consumer imaging (e.g., for panoramic image stitching), they cannot directly be used to blend images originating from NDT inspections. Indeed, in NDT images, it is necessary to retain the robustness of quantitative information (e.g., to perform pixel intensity comparisons) and avoid introducing any image processing artefacts. A typical challenge lies in removing low-frequency exposure variations while retaining sharp intensity gradients that may indicate the presence of small defects. In other words, it is necessary to prevent blurring. In this work, image blending has been solved through a method that preserves the valuable NDT information in each image. All pixel intensities in an image are offset by a unique value to maintain gradients unaltered. To explain this approach, Figure 4a,b provides an example of nine aligned images. The intensity discontinuity between any two overlapping images has been purposely emphasised. These example images do not contain high contrast features, which are typical for NDT images taken of a not-defected sample.

The idea is to shift the intensity of all pixels in an image vertically by a particular corrective value. Thus, n being the number of images in the set, it is necessary to compute a vector of n scalar optimum intensity correction values ($c^* = [c_1^*, c_2^*, \ldots c_i^*, \ldots c_n^*]$) that simultaneously correct all images in the set. These values may be positive or negative to produce an increase or a decrease in image pixel intensities. Interestingly, this computation can be formalised again through a least-squares problem of the following SSD function:

$$SSD(c) = \sum_{i=1}^{n} \sum_{j=1}^{n} \sum_{h} \left\{ [I_j(x_h) + c_j] - [I_i(x_h) + c_i] \right\}^2 \quad \text{with } (j \neq i) \text{ and } (h \in H_{i,j}), \tag{5}$$

where $H_{i,j}$ is the set of pixel indices that fall within the overlap between the aligned ith and jth image. It must be noted that the formulation of this SSD function follows the same approach used for the computation of the alignment parameters. The summation is not evaluated for $j = i$ (no intensity self-correction is required) and for combinations of i and j corresponding to images that do not overlap, where $H_{i,j}$ is an empty set. Since intensity correction is performed after the alignment stage, Equation (5) does not perform any image transformation. Moreover, since the problem is limited to the computation of only one scalar parameter per image, convergence to a solution for Equation (5) is obtained faster than for Equation (3). Once the optimum intensity correction values are found, the matrix of corrected pixel intensities for the ith image ($\widetilde{I}_i$) is computed with the following equation.

$$\widetilde{I}_i = I_i + c_i^* \qquad (6)$$

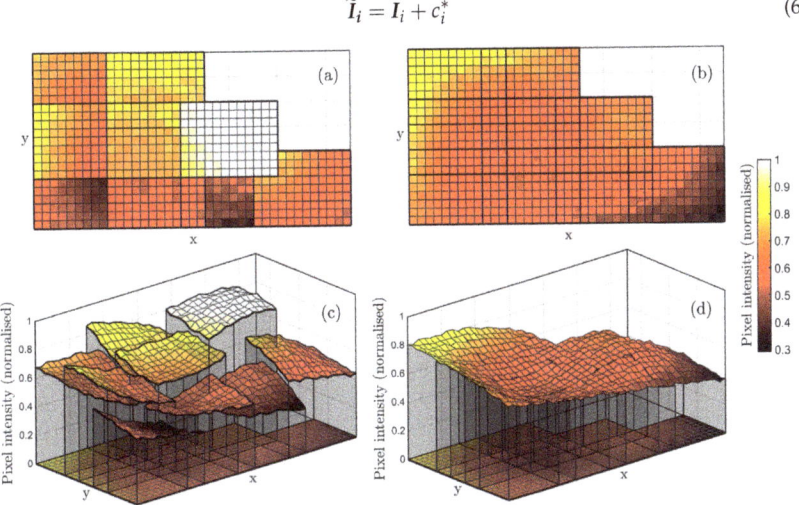

Figure 4. Exemplification of image blending, used in FiPAM. (**a**) Aligned images with discontinuous pixel intensities; (**b**) images after correction of discontinuities; (**c**) 3D plot of uncorrected pixel intensities; (**d**) 3D plot of the corrected image set.

Once all images are aligned and their intensity is corrected, the final composite image is obtained by applying the Laplacian pyramid blending, which allows a smooth transition between images. The application of blending at the end of the procedure is admissible since it does not introduce any image artefact when pixel intensity differences are low, which is the case after the phase of image pixel correction.

4. Robotic Thermography Setup

4.1. Inspection System Integration

Figure 5 illustrates the automatic thermography setup used in this work. The robotic manipulator was a KUKA KR10 R1100-2 arm [39], with a maximum payload of 11.1 kg and a maximum reach of 1101 mm. The setup was designed to perform PPT inspection in reflection mode, meaning that the flash lamp and the IR camera were always kept on the same side of the part under inspection. A custom-built supporting bracket was used to mount the flash lamp and the IR camera onto the robot and keep them in a fixed relative position during the inspection. The support allowed adjusting the orientation of the flash lamp to set the angular offset between the flash lamp illumination axis and the camera axis. This adjustment is not active because an actuator does not vary it during the execution of a robotic inspection path. However, keeping the camera focal distance constant for all data collection poses in a path makes it possible to manually set the optimum angular offset for any chosen camera focal distance before executing the inspection path. The heat source was an Elinchrom Twin X4 Lamphead EL20181, capable of releasing a pulse of 4800 W/s with a duration of 5.56 ms (1/180 s), powered by two power supplies in a parallel configuration [40]. The excitation source features a lightweight aluminium chassis, two twin flash tubes and twin cables connected to two Elinchrom 2400 RX power packs. The presence of two flash tubes and two power packs allows shorter flash durations and faster recycle times than a single flash tube connected to a single power pack, which is advantageous for the robotisation of the thermographic inspection. The IR camera was a cooled FLIR X6540sc IR-camera [41], equipped with a 50 mm F/2.0 lens; it has an adjustable acquisition rate of up to 125 Hz at full frame. The camera detector consists of 640 × 512 pixels, cooled by a Stirling thermodynamic cycle that uses an Indium-Antimonide fluid. The camera was connected to the computer through the Gigabit Ethernet link for full bandwidth data acquisition. The FLIR ResearchIR Max® software (version 4.40.1), running on the computer, enabled the initial configuration of the camera and the reception of the thermographic data during the robotic inspection. DFT was used to evaluate the frequency content of the thermal response.

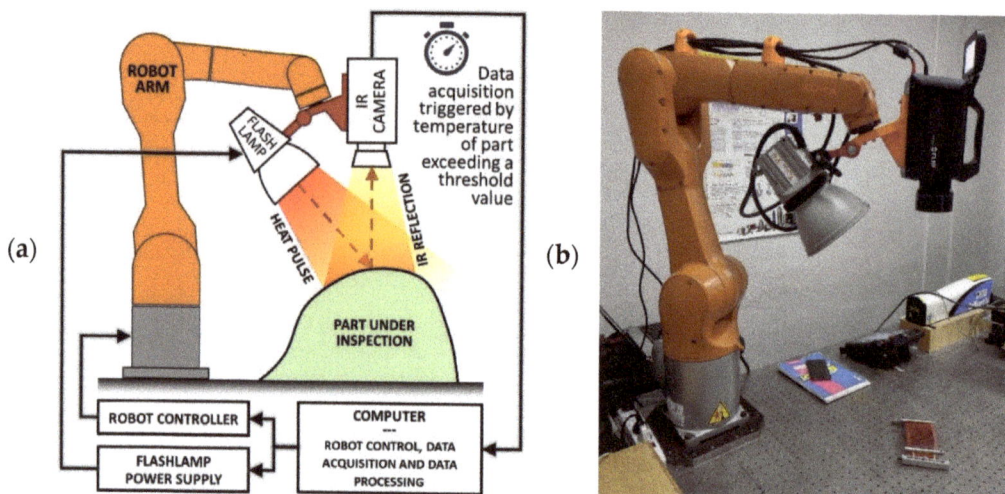

Figure 5. (a) Schematic representation of the robotic thermographic setup used in this work; (b) Photo of the actual laboratory setup.

4.2. Sample

The sample was an epoxy specimen reproducing the curved geometry of a compressor blade. The specimen was produced by pouring a mix of liquid epoxy resin and a hardener into a mould. The resultant polymerised sample had one convex side, one concave side, and a varying thickness. The curvature of both surfaces is constrained to one direction. Six flat bottom holes (FBHs), three with square sections and three with round sections, were machined on the concave side of the sample as artificial defects. Thus, the FBHs are not visible from the convex side of the sample. Figure 6a,b illustrate the sample geometry, its main dimensions, and the position and size of the FBHs. The sample was coated with acrylic-based black matt paint to uniformise and enhance the surface emissivity, improving the effectiveness of PT inspection. The sample was placed on the optical table at a registered position within the working envelope of the robot arm, using a fixed custom supporting base. The specimen was inspected from the convex side. Figure 6c shows the sample ready for inspection. In order to validate the proposed alignment method, as will become clear in the following sections, the robotic thermographic inspection was also performed by wrapping the sample with a flexible plastic 3D printed grid (as shown in Figure 6d). The grid square pattern had a 3 mm pitch and wire width of 0.6 mm.

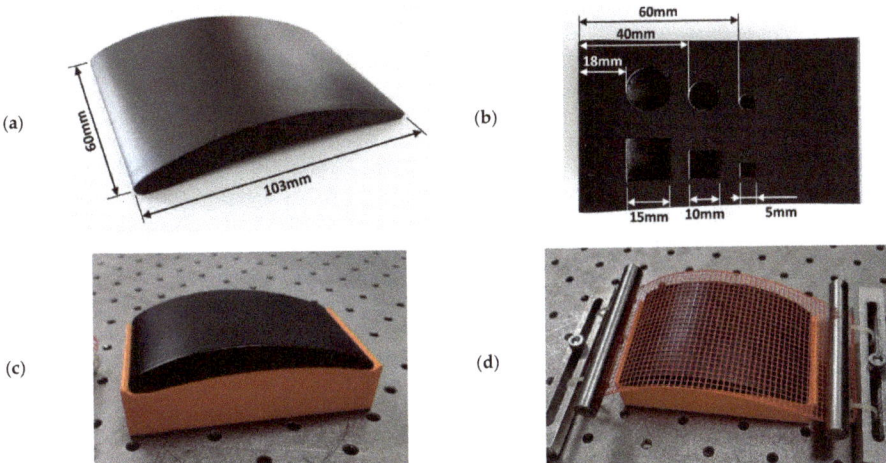

Figure 6. (**a**) Picture of the sample with the indication of footprint dimensions; (**b**) picture of the back wall with the indication of FBH locations and sizes; (**c**) sample placed on the supporting base without grid; (**d**) sample with the grid.

4.3. Robot Path-Planning, Simulation and Control

Six-axis robotic arms have traditionally been used in production lines to perform pick-and-place operations (e.g., palletising robots). In that scenario, where the exact trajectory between any two consecutive poses is not too important, a robot can be manually programmed by simply teaching the robot controller the coordinates of a few poses. Such teaching is usually performed by manually jogging the robot to each desired pose to record its coordinates. Then a robot programme is manually written to move the robot through the recorded poses. More recently, accurate mechanical joints and control units have made industrial robotic arms precise enough for finishing tasks in manufacturing operations [42]. As a result, software brands and robot manufacturers have developed many software applications to help technicians and engineers in programming complex robot tasks [43]. Using such software platforms to program robot movements is known as off-line programming (OLP). It is based on importing the 3D virtual model of the complete robot work cell, the robot end-effector, and the sample(s) to be manipulated or machined. Such robotic OLP software modules usually evolve from CAD/CAM applications, suited to programming Computer Numerical Control (CNC) manufacturing machines.

Despite the abundance of OLP software solutions geared towards manufacturing applications, limited solutions have been demonstrated for robotic NDT delivery [44,45]. Using commercial OLP software to generate appropriate tool paths for NDT purposes may seem relatively straightforward at first glance, but there are several inadequacies:

- Many commercial software applications for robotic off-line programming are expensive tools, incorporating a lot of functionality specific for CAD/CAM purposes and machining features;
- Path-planning for automated NDT inspections is a very particular task. Conventional OLP software has no accessible provision for tool-path customisation to accommodate the requirements of NDT inspections;
- Commercially available OLP software does not provide capabilities for full synchronisation between robotic movements and NDT data acquisition from sensor instrumentation systems (e.g., the thermographic IR camera, in this case). Such synchronisation is fundamental to enable the possibility of positional encoding of the NDT data to create accurate NDT maps of an inspected part [45].

In this work, robotic path-planning, simulation and control for automated thermographic inspection were enabled through developing a bespoke MATLAB-based graphic user interface. Figure 7 shows a screenshot of the application taken during the path-planning phase to inspect the sample described above. This software application imports the digital models of the robot, the thermographic instrumentation and the sample, producing a virtual representation of the inspection setup. The application was mainly developed to enable the automated thermographic data collection required for validating the data alignment method introduced by this work. Although it has no ambition to be a fully-developed software tool, it contains vital features to allow flexibility and future usability. The digital sample model is positioned in the virtual scene according to the user-specified coordinates for the sample reference frame with respect to the robot reference system. The set of coordinates comprises the three Cartesian coordinates of the sample origin and the three Eulerian angular coordinates of the coordinated axes. Although the application does not allow easy replacement of the employed thermographic instrumentation, provision has been made to enable customisation of the IR camera focal distance. Indeed, the inspection resolution depends on the camera's distance from the sample surface for a given camera lens with a fixed focal distance. Thus, changing the camera focal distance is greatly important to allow accurate planning and simulation of the robotic task. The indication of the camera focal distance enables the software to compute the robot tool centre point (TCP) coordinates. The application allowed the creation of a raster inspection tool-path for the sample, according to the user-specified maximum spacing between consecutive image acquisition poses (25 mm) and offset from the sample edges (10 mm), resulting in an inspection path consisting of 15 data acquisition poses arranged in three passes (5 poses per pass). The TCP is kept on the part surface for all poses. The z-direction of the tool reference frame follows the surface's normal direction to keep the camera view axis always perpendicular to the surface. Due to the curvature of the surface, the fact that the IR camera view axis is kept perpendicular to the surface does not guarantee that all the infrared rays emitted by the surface are perpendicular to the camera sensor. That aspect can be neglected by reducing the part surface area imaged from a single camera position, which is the main reason for employing robotic thermography. The surface area imaged from each camera position is reduced by bringing the camera closer to the part and/or cropping the camera's full image frame (sub-windowing). The sub-windowing also allows higher frame acquisition rates, resulting in a better temporal sampling of the thermal wave.

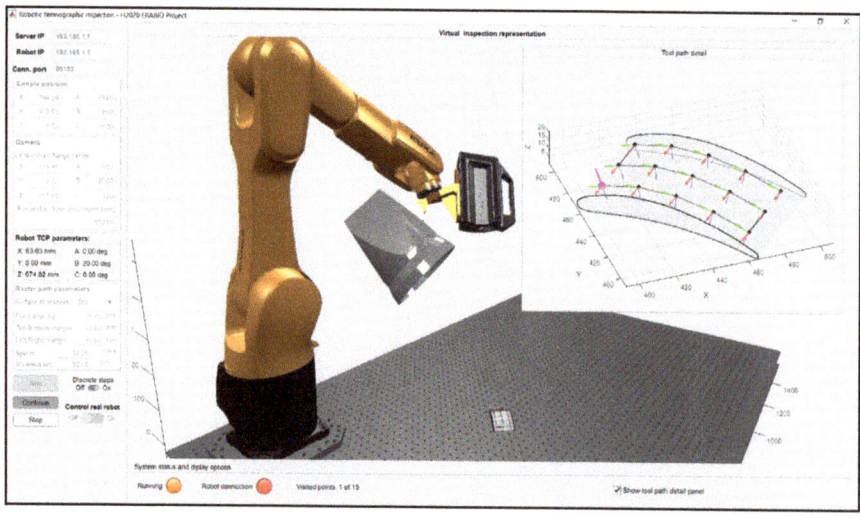

Figure 7. MATLAB-based graphic user interface for robot path-planning, simulation and control.

The application allows simulating the automated task workflow before sending the path command coordinates to the connected robot. The connection between the computer and the robot was managed through the Interfacing Toolbox for Robotic Arms (ITRA) [46]. The ITRA allowed synchronising the robotic camera manipulation and the data collection to carry out the following steps, supported by the schematic representation given in Figure 5a:

1. The computer sends the command coordinates of one inspection pose to the robot controller and waits for a digital acknowledgement from it, which signals the arrival of the robot arm at the commanded pose;
2. While the robot is at a standstill, the flashlamp power supply is triggered;
3. In turn, the sample surface temperature rise resulting from the flashlamp heat pulse triggers the IR camera data acquisition;
4. The computer acquires the raw camera data through the FLIR ResearchIR Max® software;
5. The previous steps repeat for the following inspection pose until all poses are visited.

Figure 8 shows the robotic inspection system at the first five path poses. A video of the robotic data acquisition is available for download as Supplementary Material.

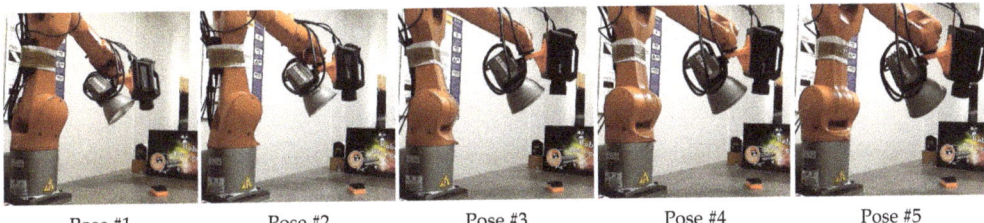

Figure 8. Robotic inspection system during data acquisition for the first five poses.

5. Results

Figures 9 and 10 show the sets of thermographic images acquired with the tool path presented in Section 4.3, using a camera focal distance of 550 mm. The camera acquired the evolution of the thermographic field for 10 s at each pose (starting from one second before the trigger of the flashlamp). DFT was used to evaluate the frequency content of the thermal response at 0.6 Hz. All images have the same size (192 × 224 pixels). They relate to the images captured from the sample with and without the grid. Thus, the same robotic tool path was repeated twice to ensure repeatability in the acquisition poses. It must be noted how the average pixel intensity varies from image to image within each set; for example, image #4 and image #12 respectively present significant lower and higher average intensity than the rest of the images in the first set, respectively. Furthermore, pixel intensity is not repeatable since differences are evident across the two sets. Any pair of corresponding images in the two sets present a visible difference in pixel intensity.

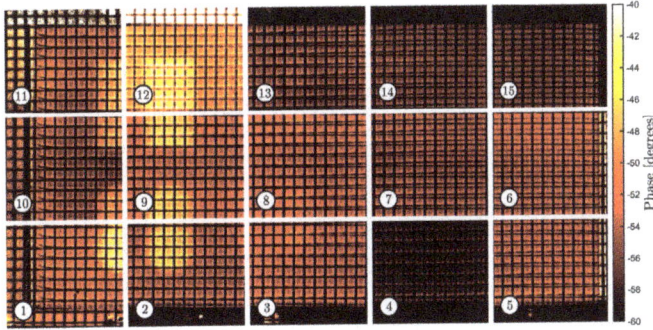

Figure 9. Set of phasegrams taken from the sample with the superposed grid.

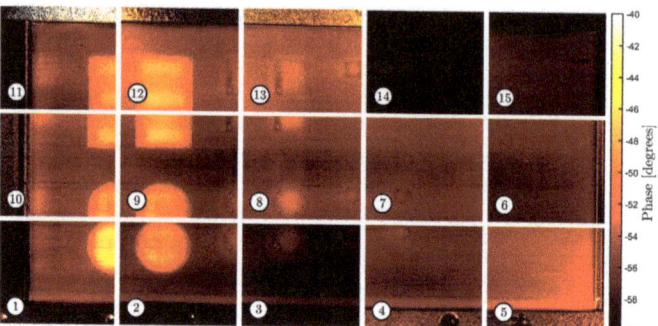

Figure 10. Set of phasegrams taken from the sample without the superposed grid.

Figure 11 highlights the initial estimate of the overlaps in each set of images. The images were encoded with the camera shooting positions and scaled by the measured resolution value. The known pitch of the grid (3 mm) was used to estimate the resolution of the images, which was 150 μm/pixel ($\cong$4444 pixels/cm^2). Figure 11a,b relate to the set of images with and without the grid, respectively. There, the image pixels were plotted with 50% transparency to allow visualising the overlaps, which are more clearly illustrated in Figure 11c. Following the notation introduced in Section 3.3, the presented FiPAM method was employed using a value of 1 mm for the offset (o) between the image overlap areas and the shrunk areas. This equates to assuming that the maximum distance between a pair of corresponding pixels in neighbour images (the misalignment) does not exceed 1 mm, which is the case for the sets of images at hand.

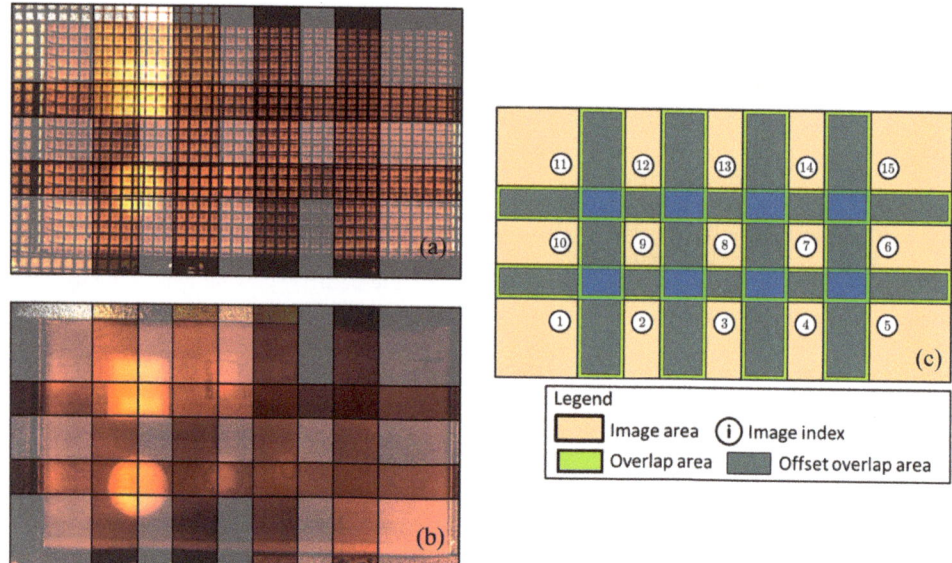

Figure 11. Plots of scaled and encoded images. (**a**) Set of images taken from the sample with the grid; (**b**) Images taken from the sample without the grid; (**c**) Overlapping areas.

As stated above, the FiPAM algorithm contains the mathematical parametric formulation of all five typical 2D image transformations (translation, Euclidean, similarity, affinity and homography), allowing aligning all images in a set with the same type of planar transformation or using a different type of transformation for each image. Each image can be given a diverse set of DoFs and treated in six different ways if no transformation (no degrees of freedom) is included as an additional option, corresponding to a total of $6^{15} \cong 4.70 \cdot 10^{11}$ possible diverse ways of applying FiPAM to our sets of fifteen images.

Figures 12 and 13 illustrate the results obtained with FiPAM, using the similarity transformation for all images. The similarity transformation, which allows four DoFs (horizontal translation, vertical translation, rotation and scaling), proved sufficient to permit a fine alignment of all images in the given sets. In Figures 12a and 13a, whereas the dotted blue line rectangles represent a fivefold scaled-down version of the original images, the rectangles with a green line perimeter represent a twofold scaled-down version of the aligned images, where the translation is magnified by a factor of 20 and the rotation transformation is maximised by a factor of 40. These magnifications were introduced to illustrate the computed alignment transformations for visualisation purposes. The infill colour given to each aligned image is linked to the computed pixel intensity correction through the indicated colour map. The resulting blended mosaic thermographic image (460 × 800 pixels) is given in Figures 12b and 13b for the images with and without the grid, respectively.

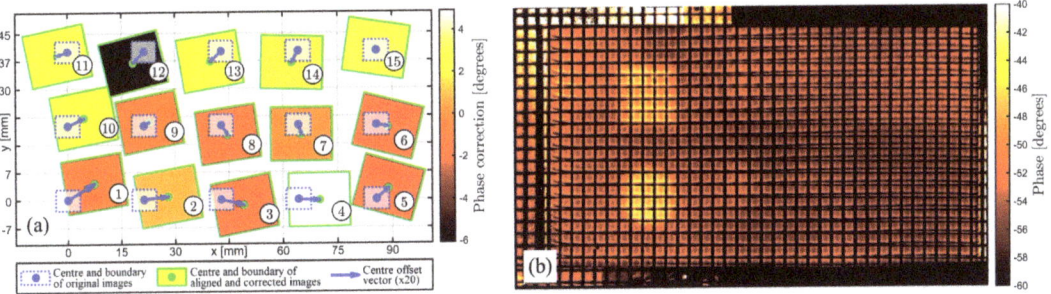

Figure 12. (**a**) Schematic illustration of similarity transformations and pixel intensity corrections computed through the proposed method for the set of images relative to the sample with the grid. (**b**) Resulting composite mosaic thermographic image.

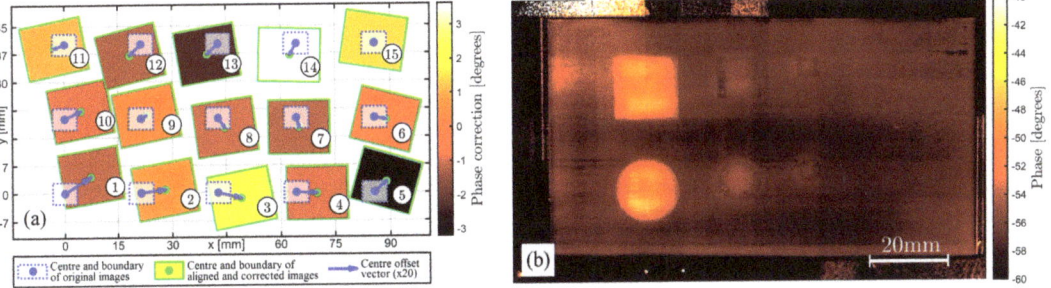

Figure 13. (**a**) Schematic illustration of similarity transformations and pixel intensity corrections computed through the proposed method for the set of images relative to the sample without the grid. (**b**) Resulting composite mosaic thermographic image.

Although testing FiPAM with all transformation combinations is not viable, the method was evaluated through a representative subset by employing each of the five possible planar transformations for all images and changing the number of images to align; this allowed varying the problem size significantly to evaluate the execution time of FiPAM. The number of images considered for each type of transformation was: 2, 5, 10 and 15, corresponding to aligning the first two images, the five images collected in the first pass of the tool path, the images in the first two passes or all the images in the set. As a result, the total number of DoFs considered in the alignment problem spanned from four, for two images transformed through pure translation (two parameters per image), to 120, for fifteen images transformed through homography (eight parameters per image). FiPAM was implemented and evaluated through MATLAB (version 2020b), running on a computer with an Intel® i7-6820HQ CPU (2.70 GHz, 4 Cores) and 32 Gb of Random-Access Memory. The MATLAB implementation code developed in this work is accessible at https://doi.org/10.5281/zenodo.6817052 (accessed on 11 July 2022). The recorded execution times are plotted in Figure 14.

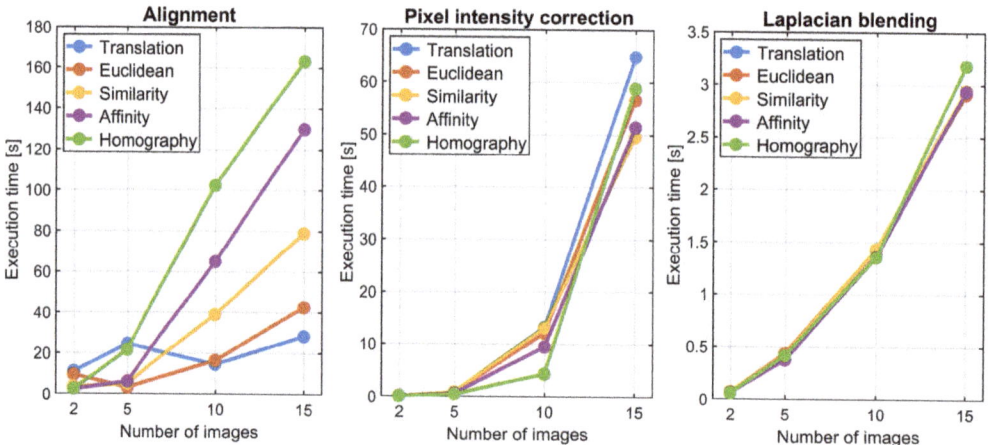

Figure 14. Execution times for alignment, pixel intensity correction and Laplacian blending.

6. Discussion

Traditional manual inspection approaches are insufficient in some scenarios. Therefore, there are fundamental motivations for increasing automation in non-destructive testing. Automation of NDT is required to cope with the inspection of large and/or curved geometries. The key challenges to face when developing a robotic NDT system include integrating the NDT instrumentation with the robotic manipulator, creating a suitable robot inspection path for the part under inspection, and developing software for NDT data collection and visualisation. Although these challenges have been addressed by several applications of six-axis robotic arms for the inspection of parts through automated ultrasonic techniques, other types of inspections have not reached the same level of robotisation, which is the case with thermographic testing. This work bridges a technology gap, making thermographic inspections more deployable in industrial environments. Furthermore, the proposed fine image alignment method (FiPAM) can find applicability beyond thermographic NDT.

The results prove that FiPAM enables the proper merging of multiple thermographic images into one single mosaic image, which is easier to analyse. This is accomplished through three steps: simultaneous alignment of all images in a set, global optimum pixel intensity correction, and image blending. The reported composite mosaic images, in Figures 12b and 13b, obtained through computing similarity transformations and pixel corrections for the images acquired in this work, show a significant reduction of the original discontinuities. Whereas the scale of the composite image relative to the sample with the grid is immediately retrievable from the known grid pitch (3 mm), a reference 20 mm long scale bar was added to the image relative to the sample without the grid. It is straightforward to note that the sizes of the thermographic indications correspond to the physical sizes of the artificial FBHs. The difference in thermographic pixel intensity for FBHs of diverse sizes is coherent with the change in the aspect ratio between heat blocking and leakage surface, as described in [47]. Larger FBH diameter to depth ratios produce the emergence of localised higher intensities in the IR image sequence.

Although FiPAM execution times are machine-dependent, the patterns presented in Figure 14 provide a helpful guideline for understanding the general trends. As expected, the execution times for alignment, pixel intensity correction and Laplacian blending increase with the number of images. The alignment phase execution time also depends on the type of transformations used for the images in the set. They influence the size of the least-squares problem and the number of transformation parameters to find through the minimisation of the SSD function in Equation (3). Thus, for a given number of images to align, using the same type of transformation for all images, the execution time increases monotonically, moving from translation to Euclidean, similarity, affinity and homography transformations. Although all possible combinations are not assessed in this work, it is not difficult to imagine intermediate execution times for generic combinations, where not all the images get transformed by the same transformation type. As a rule of thumb, for a given number of images, the alignment execution time should never exceed the time relative to the case where all images get transformed through homography, since it corresponds to the biggest problem with the maximum number of parameters. Fluctuations in alignment execution times can be observed for patterns relative to translation and Euclidean transformations. They are thought to be caused by the limited DoFs allowed by these transformations, which can cause prolonged convergence times due to the difficulty of obtaining a good image alignment. The execution times of the pixel correction and Laplacian blending phases depend on the number of images. The minor differences associated with the used transformation type are thought to be caused by the different overlaps of the aligned images, which changes the number of pixel intensity differences to compute for the SSD function in Equation (4).

The advantages of FiPAM, described in this work, should be clear by now. One limitation of the current implementation is that FiPAM is suitable for aligning multiple images of a sample surface that curves only in one direction. Although this limitation does not impede using FiPAM for generalised cylindrical surfaces, future work should focus on extending FiPAM to operate with images encoded in three-dimensional space.

Supplementary Materials: The following supporting information can be downloaded at: https://www.mdpi.com/article/10.3390/s22166267/s1, Video S1: Robotic thermographic data acquisition.

Author Contributions: Conceptualisation, C.M., N.M. and A.P.; methodology, C.M. and N.M.; software, C.M.; validation, C.M., N.M. and M.F.; formal analysis, C.M.; investigation, N.M. and M.F.; resources, C.M. and N.M.; data curation, C.M.; writing—original draft preparation, C.M.; writing—review and editing, all; visualisation, C.M.; supervision, D.C. and A.P.; project administration, D.C. and A.P.; funding acquisition, C.M. and D.C. All authors have read and agreed to the published version of the manuscript.

Funding: This work received funding from the European Union's Horizon 2020 Research and Innovation Programme under the Marie Sklodowska-Curie grant agreement No. 835846.

Institutional Review Board Statement: Not applicable.

Informed Consent Statement: Not applicable.

Data Availability Statement: The source code for the MATLAB-based implementation of FiPAM and the example thermographic dataset are available at https://doi.org/10.5281/zenodo.6817052 (accessed on 11 July 2022).

Conflicts of Interest: The authors declare no conflict of interest. The funders had no role in the design of the study; in the collection, analyses, or interpretation of data; in the writing of the manuscript, or in the decision to publish the results.

Abbreviations

The following abbreviations are used in this manuscript:

CAD	Computer-Aided Design
CAM	Computer-Aided Manufacturing
CNC	Computer Numerical Control
CPU	Central Processing Unit
DFT	Discrete Fourier Transform
FBH	Flat Bottom Hole
FiPAM	Fine Pixel-based Alignment Method
IR	Infrared
ITRA	Interfacing Toolbox for Robotic Arms
LT	Lock-in Thermography
NDT	Non-Destructive Testing
OLP	Off-Line Programming
PPT	Pulsed Phase Thermography
PT	Pulsed Thermography
RGB	Red, Green and Blue
SSD	Sum of Squared Differences
TCP	Tool Centre Point

References

1. Hull, J.B.; John, V. *Non-Destructive Testing*; Macmillan International Higher Education: London, UK, 2015.
2. Sattar, T.P. Robotic Non-Destructive Testing. *Ind. Robot. Int. J.* **2010**, 37. [CrossRef]
3. Gibson, I.; Rosen, D. *Brent Stucker, and Mahyar Khorasani. Additive Manufacturing Technologies*; Springer: Berlin/Heidelberg, Germany, 2021; Volume 17.
4. Appleton, E.; Williams, D.J. *Industrial Robot Applications*; Springer Science & Business Media: Berlin/Heidelberg, Germany, 2012.
5. Mineo, C.; Pierce, S.G.; Wright, B.; Cooper, I.; Nicholson, P.I. PAUT inspection of complex-shaped composite materials through six DOFs robotic manipulators. *Insight Non-Destr. Test. Cond. Monit.* **2015**, *57*, 161–166. [CrossRef]
6. Cuevas, E.; Lopez, M.; García, M.; Ibérica, K.R. Ultrasonic Techniques and Industrial Robots: Natural Evolution of Inspection Systems. In Proceedings of the 4th International Symposium on NDT in Aerospace, Berlin, Germany, 13–15 November 2012.
7. Bosse, J.; Thaler, B.; Ilse, D.; Bühling, L. Automated Air-Coupled Ultrasonic Technique for the Inspection of the Ec145 Tail Boom. Available online: http://2012.ndt-aerospace.com/Portals/aerospace2012/BB/tu2b3.pdf (accessed on 11 July 2022).
8. Mineo, C.; Riise, J.; Summan, R.; MacLeod, C.N.; Pierce, S. Index-based triangulation method for efficient generation of large three-dimensional ultrasonic C-scans. *Insight Non-Destr. Test. Cond. Monit.* **2018**, *60*, 183–189. [CrossRef]
9. Almadhoun, R.; Taha, T.; Seneviratne, L.; Dias, J.; Cai, G. A survey on inspecting structures using robotic systems. *Int. J. Adv. Robot. Syst.* **2016**, *13*, 1729881416663664. [CrossRef]
10. Meola, C. *Infrared Thermography Recent Advances and Future Trends*; Bentham Science Publishers: Bussum, The Netherlands, 2012.
11. Holst, G.C. *Common Sense Approach to Thermal Imaging*; SPIE Optical Engineering Press: Washington, WA, USA, 2000; Volume 1.
12. Maldague, X. *Theory and Practice of Infrared Technology for Nondestructive Testing*; John Wiley & Son: New York, NY, USA, 2001.
13. Schmidt, T.; Dutta, S. *Automation in Production Integrated Ndt Using Thermography*; NDT Aerospace: Augsburg, Germany, 2012; p. 8.
14. Massaro, A.; Galiano, A. Infrared Thermography for Intelligent Robotic Systems in Research Industry Inspections: Thermography in Industry Processes. In *Handbook of Research on Advanced Mechatronic Systems and Intelligent Robotics*; IGI Global: Hershey, PA, USA, 2020; pp. 98–125.
15. Holland, S.D.; Krishnamurthy, A. Registration of Nde Data to Cad. In *Handbook of Nondestructive Evaluation 4.0*; Springer: Berlin/Heidelberg, Germany, 2022; pp. 369–402.
16. Maldague, X.; Moore, P.O. *Non-Destructive Handbook, Infrared and Thermal Testing*; ASNT Press: Columbus, OH, USA, 2001; Volume 3.
17. Martin, R.E.; Gyekenyesi, A.L.; Shepard, S.M. Interpreting the Results of Pulsed Thermography Data. *Mater. Eval.* **2003**, *61*, 611–616.
18. Balageas, D.L. In Search of Early Time: An Original Approach in the Thermographic Identification of Thermophysical Properties and Defects. *Adv. Opt. Technol.* **2013**, *2013*, 314906. [CrossRef]
19. Ibarra-Castanedo, C.; Genest, M.; Servais, P.; Maldague, X.P.V.; Bendada, A. Qualitative and quantitative assessment of aerospace structures by pulsed thermography. *Nondestruct. Test. Eval.* **2007**, *22*, 199–215. [CrossRef]
20. Maldague, X.; Marinetti, S. Pulse phase infrared thermography. *J. Appl. Phys.* **1996**, *79*, 2694–2698. [CrossRef]
21. Maldague, X.; Galmiche, F.; Ziadi, A. Advances in pulsed phase thermography. *Infrared Phys. Technol.* **2002**, *43*, 175–181. [CrossRef]
22. Ibarra-Castanedo, C.; Maldague, X. Pulsed Phase Thermography Reviewed. *Quant. Infrared Thermogr. J.* **2004**, *1*, 47–70.

23. BuSSe, G.; Wu, D.; Karpen, W. Thermal Wave Imaging with Phase Sensitive Modulated Thermography. *J. Appl. Phys.* **1992**, *71*, 3962–3965. [CrossRef]
24. Mineo, C.; Cerniglia, D.; Poole, A. Autonomous Robotic Sensing for Simultaneous Geometric and Volumetric Inspection of Free-Form Parts. *J. Intell. Robot. Syst.* **2022**, *105*, 54. [CrossRef]
25. Zhang, Q.; Tian, W.; Hu, J.; Li, P.; Wu, C. Robot Hand-Eye Calibration Method Based on Intermediate Measurement System. In Proceedings of the International Conference on Intelligent Robotics and Applications 2021, Yantai, China, 22–25 October 2021.
26. Schramm, S.; Osterhold, P.; Schmoll, R.; Kroll, A. Combining modern 3D reconstruction and thermal imaging: Generation of large-scale 3D thermograms in real-time. *Quant. Infrared Thermogr. J.* **2021**, 1–17. [CrossRef]
27. Paragios, N.; Chen, Y.; Faugeras, O.D. *Handbook of Mathematical Models in Computer Vision*; Springer Science & Business Media: Berlin/Heidelberg, Germany, 2006.
28. Szeliski, R. Image alignment and stitching: A tutorial. *Found. Trends Comput. Graph. Vis.* **2007**, *2*, 1–104.
29. Keysers, D.; Deselaers, T.; Ney, H. Pixel-to-Pixel Matching for Image Recognition Using Hungarian Graph Matching. In *Joint Pattern Recognition Symposium*; Springer: Berlin/Heidelberg, Germany, 2004; pp. 154–162. [CrossRef]
30. Zoghlami, I.; Faugeras, O.; Deriche, R. Using Geometric Corners to Build a 2d Mosaic from a Set of Images. In Proceedings of the IEEE Computer Society Conference on Computer Vision and Pattern Recognition 1997, San Juan, PR, USA, 17–19 June 1997.
31. Cham, T.-J.; Cipolla, R. A statistical framework for long-range feature matching in uncalibrated image mosaicing. In Proceedings of the 1998 IEEE Computer Society Conference on Computer Vision and Pattern Recognition (Cat. No. 98CB36231), Santa Barbara, CA, USA, 25–25 June 1998; IEEE: Manhattan, NY, USA, 2002. [CrossRef]
32. Brown, M.; Lowe, D.G. Recognising Panoramas. In Proceedings of the 9th IEEE International Conference on Computer Vision (ICCV 2003), Nice, France, 14–17 October 2003.
33. Meng, X.; Wang, Y.; Liu, J.; He, W. Non-destructive Inspection of Curved Clad Composites with Subsurface Defects by Combination Active Thermography and Three-Dimensional (3d) Structural Optical Imaging. *Infrared Phys. Technol.* **2019**, *97*, 424–431. [CrossRef]
34. Albert, A.A. *Solid Analytic Geometry*; Courier Dover Publications: Mineola, NY, USA, 2016.
35. Fan, Q.; Zhang, L. A novel patch matching algorithm for exemplar-based image inpainting. *Multimedia Tools Appl.* **2017**, *77*, 10807–10821. [CrossRef]
36. Fadnavis, S. Image Interpolation Techniques in Digital Image Processing: An Overview. *Int. J. Eng. Res. Appl.* **2014**, *4*, 70–73.
37. Burt, P.J.; Adelson, E.H. A multiresolution spline with application to image mosaics. *ACM Trans. Graph.* **1983**, *2*, 217–236. [CrossRef]
38. Levin, A.; Zomet, A.; Peleg, S.; Weiss, Y. Seamless Image Stitching in the Gradient Domain. In Proceedings of the European Conference on Computer Vision 2004, Prague, Czech Republic, 11–14 May 2004.
39. KUKA. Kr 10 R1100-2. Available online: https://www.kuka.com/-/media/kuka-downloads/imported/6b77eecacfe542d3b736af377562ecaa/0000290003_en.pdf?rev=3e82b095d46c4a86b7e195cabdb980cb&hash=A4F0F7D35D9485B612F8BC455B5805E4 (accessed on 16 June 2022).
40. Elinchrom. Digital Rx. Available online: https://www.elinchrom.com/wp-content/uploads/download-center/73256_digital_rx_manuel--en-de-fr.pdf (accessed on 16 June 2022).
41. FLIR. Flir X6540sc. Available online: http://www.flir.at/fileadmin/user_upload/Vertretungen/FLIR/X6900sc/Datasheets/X6540sc_66701-0101-en-US_A4-1706-nbn.pdf (accessed on 16 June 2022).
42. Bogue, R. Finishing robots: A review of technologies and applications. *Ind. Robot. Int. J. Robot. Res. Appl.* **2009**, *36*, 6–12. [CrossRef]
43. Pan, Z.; Polden, J.; Larkin, N.; Van Duin, S.; Norrish, J. Recent progress on programming methods for industrial robots. *Robot. Comput. Manuf.* **2012**, *28*, 87–94. [CrossRef]
44. Haase, W.; Ungerer, D.; Mohr, F. Automated Non-Destructive Examination of Complex Shapes. In Proceedings of the 14th Asia-Pacific Conference on NDT (APCNDT), Mumbai, India, 18–22 November 2013.
45. Mineo, C.; Pierce, S.G.; Nicholson, P.I.; Cooper, I. Robotic path planning for non-destructive testing—A custom MATLAB toolbox approach. *Robot. Comput. Manuf.* **2016**, *37*, 1–12. [CrossRef]
46. Mineo, C.; Vasilev, M.; Cowan, B.; MacLeod, C.N.; Pierce, S.G.; Wong, C.; Yang, E.; Fuentes, R.; Cross, E.J. Enabling Robotic Adaptive Behaviour Capabilities for New Industry 4.0 Automated Quality Inspection Paradigms. *Insight-Non-Destr. Test. Cond. Monit.* **2020**, *62*, 338–344.
47. Beemer, M.F.; Shepard, S.M. Aspect ratio considerations for flat bottom hole defects in active thermography. *Quant. Infrared Thermogr. J.* **2017**, *15*, 1–16. [CrossRef]

Article

Defects Recognition Algorithm Development from Visual UAV Inspections

Nicolas P. Avdelidis [1,*], Antonios Tsourdos [1], Pasquale Lafiosca [1], Richard Plaster [2], Anna Plaster [2] and Mark Droznika [3]

1. School of Aerospace, Transport & Manufacturing, Cranfield University, Cranfield MK43 0AL, UK; a.tsourdos@cranfield.ac.uk (A.T.); pasquale.lafiosca@cranfield.ac.uk (P.L.)
2. ADDIT, 17 Railton Road, Wolseley Business Park, Kempston, Bedford MK42 7PN, UK; richardjamesplaster@gmail.com (R.P.); annaplaster@gmail.com (A.P.)
3. TUI Airline, Area 8, Hangar 61, Percival Way, London Luton Airport, Luton LU2 9PA, UK; mark.droznika@tui.co.uk
* Correspondence: np.avdel@cranfield.ac.uk; Tel.: +44-(0)1234-754366

Citation: Avdelidis, N.P.; Tsourdos, A.; Lafiosca, P.; Plaster, R.; Plaster, A.; Droznika, M. Defects Recognition Algorithm Development from Visual UAV Inspections. *Sensors* **2022**, *22*, 4682. https://doi.org/10.3390/s22134682

Academic Editors: Yashar Javadi and Carmelo Mineo

Received: 28 May 2022
Accepted: 20 June 2022
Published: 21 June 2022

Publisher's Note: MDPI stays neutral with regard to jurisdictional claims in published maps and institutional affiliations.

Copyright: © 2022 by the authors. Licensee MDPI, Basel, Switzerland. This article is an open access article distributed under the terms and conditions of the Creative Commons Attribution (CC BY) license (https://creativecommons.org/licenses/by/4.0/).

Abstract: Aircraft maintenance plays a key role in the safety of air transport. One of its most significant procedures is the visual inspection of the aircraft skin for defects. This is mainly carried out manually and involves a high skilled human walking around the aircraft. It is very time consuming, costly, stressful and the outcome heavily depends on the skills of the inspector. In this paper, we propose a two-step process for automating the defect recognition and classification from visual images. The visual inspection can be carried out with the use of an unmanned aerial vehicle (UAV) carrying an image sensor to fully automate the procedure and eliminate any human error. With our proposed method in the first step, we perform the crucial part of recognizing the defect. If a defect is found, the image is fed to an ensemble of classifiers for identifying the type. The classifiers are a combination of different pretrained convolution neural network (CNN) models, which we retrained to fit our problem. For achieving our goal, we created our own dataset with defect images captured from aircrafts during inspection in TUI's maintenance hangar. The images were preprocessed and used to train different pretrained CNNs with the use of transfer learning. We performed an initial training of 40 different CNN architectures to choose the ones that best fitted our dataset. Then, we chose the best four for fine tuning and further testing. For the first step of defect recognition, the DenseNet201 CNN architecture performed better, with an overall accuracy of 81.82%. For the second step for the defect classification, an ensemble of different CNN models was used. The results show that even with a very small dataset, we can reach an accuracy of around 82% in the defect recognition and even 100% for the classification of the categories of missing or damaged exterior paint and primer and dents.

Keywords: defect recognition; aircraft inspection; deep learning; CNN; UAV; defect classification; AI

1. Introduction

Air transport is one of the most significant ways of moving people across the globe. In 2019, the number of air passengers carried worldwide was around 4.2 billion, an overall increase of 92% compared with 2019 [1]. During COVID-19, most travelling was put almost on a halt with the numbers decreasing significantly. In 2020, the total number of passengers dropped significantly to around one billion (1034 million) [2]. As a result, the need of reducing costs across the industry has become imminent. Around 10–15% of the operational costs of an airline are around maintenance, repairs, and overhaul (MRO) activities [3]. Currently, aircraft maintenance heavily involves visual tasks carried by humans [3]. This is very time consuming, costly and introduces possibilities for human errors. It is understood that automating these visual tasks could solve this problem [4–6]. For this reason, the use of climbing robots or UAVs to perform these tasks have been attempted. Climbing robots usually use magnetic forces, suction caps, or vortexes to climb to the aircraft structure [7–9]. However, robotic platforms for inspection face difficulties in achieving good adherence and mobility due to their lack of flexibility [7,10,11]. On the other hand, UAVs have been proposed for the inspection [12–15] of buildings, wind turbines, power transmission lines and aircrafts. UAVs could minimize inspection time and cost as they can inspect quickly large areas of the aircraft and data can be transmitted to a ground base in real time for processing. The key challenge of all the above automated techniques is developing defect detection algorithms that are able to perform with accuracy and repeatability. Several attempts have been made and most of them can be divided into the following two categories: the ones that use more traditional image processing techniques [5,16–18] and the ones that use machine learning [19–24]. In the first category, image features such as convexity or signal intensity [5] are used. In [18], the authors proposed a method using histogram comparisons or structural similarity. In addition, in [16,17], the authors proposed the use of neural networks trained on feature vectors extracted from contourlet transform. These techniques have very good accuracy in the test data but are failing to effectively generalize and need continuous tuning. On the other hand, algorithms using convolutional neural networks (CNN) have showed good results in defect detection [19–21,25]. In [19,20], CNNs are used as feature extractors and then either a single shot multibox detector (SSD) or a support vector machine (SVM) are used for the classification. The use of faster CNN is also proposed for classification and localization [22]. In addition, the use of UAVs together with deep learning algorithms is proposed for the tagging and localization of concrete cracks [26,27].

The main challenge of the machine learning algorithms is the requirement of a large amount of data. Especially for the CNNs, the amount of data required can be in the scale of thousands, especially if a model is not already pretrained. The existence of large datasets in concrete structures has allowed CNNs to show excellent results in defect detection in concrete structures. On the other hand, in aircraft structures, the results are promising but are still not very accurate [18] or they deal with only the problem of defect recognition [20]. In this paper, we propose a two-step classification process of an ensemble of machine learning classifiers for both defect recognition and classification. In this two-step process, we are using pretrained CNNs to both recognize and classify a series of defects in aircraft metallic and composite structures. In the first step, we are performing the defect recognition and in the second step, the defect classification. By combining the results of different classifiers, we can more effectively address the issue of small datasets and produce results with an accuracy reaching 82% in the defect recognition step.

2. Dataset

One of the challenges in this study was the creation of datasets for training and testing the classifiers. As most of the datasets of defects on aircrafts are not public available, we needed to create our own. The datasets were created with the help and permission of TUI© [28]. The images were taken during the scheduled maintenance of aircrafts in TUI's base maintenance hangar in Luton, UK. The imaging sensor used was a SONY RX0 II© rugged mini camera. This model can be carried by a drone and is able to take images from any angle. All the technical specifications of the camera, such as sensor size and type, focal length, size of the sensor, are widely available and the effective resolution is 15-megapixels with maximum resolution of 4800 × 3200 pixels. Images for the datasets were captured and the following seven types of defects were investigated:

- Missing or damaged exterior paint and primer;
- Dents;
- Lighting strike damage;
- Lighting strike fastener repair;
- Reinforcing ratch repairs;
- Nicks, scratches, and gouges;
- Blend/rework repairs.

In Figure 1, images of the defects are presented. Most of the obtained images contained several defects, together with other elements such as screws, etc. In order to create the two different datasets, further processing was needed to extract only the objects that we were interested in from each of the images.

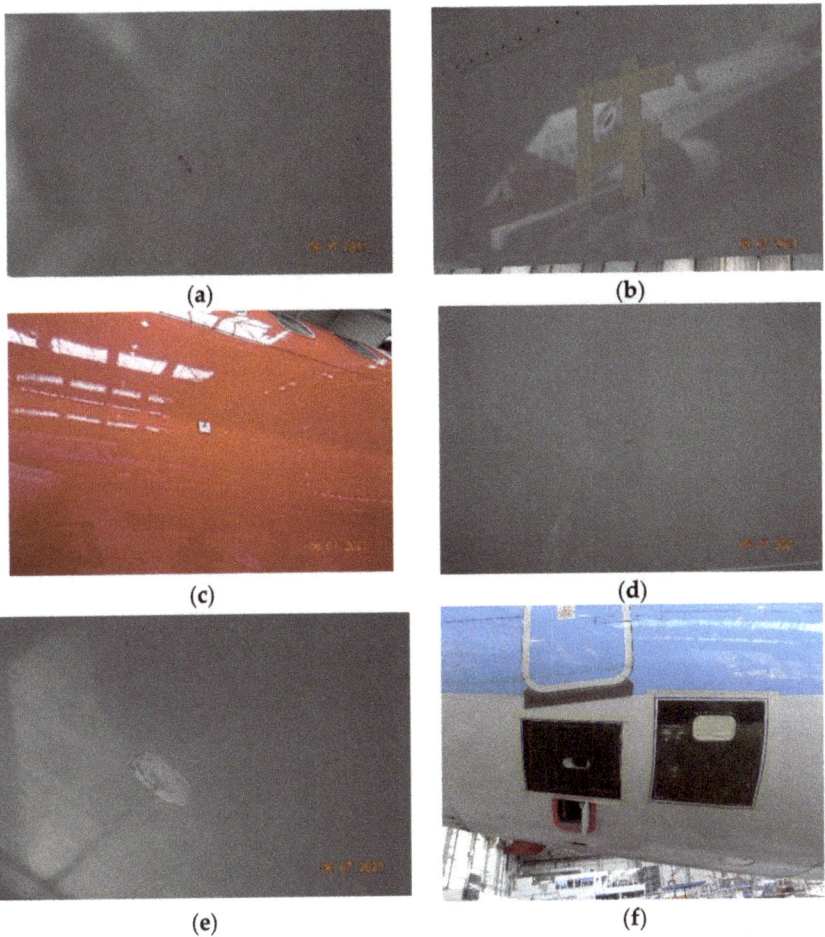

Figure 1. Images of different types of defects in aircraft structures. (**a**) Missing paint, (**b**) dents, (**c**) lighting strike damage, (**d**) lighting strike fastener repair, (**e**) blend/rework repair (material removed and then re-protected with exterior paint); (**f**) double patch repair.

The objects of interest were cropped through a semi-automated procedure to create the datasets for the training. A Python script was developed so the user can select and crop the area with the object of interest. The cropped image was saved in the new image file. The name of the file was indicative of the category of the defect. This provided us the capability to extract multiple images of interest from only one image, with and without defects. The cropped images were grayscaled because we did not want the classifiers to associate color with any defects during training. This was carried out because defects are not color related and aircrafts are painted in different colors, depending on the company. Images of the datasets can be observed in Figure 2.

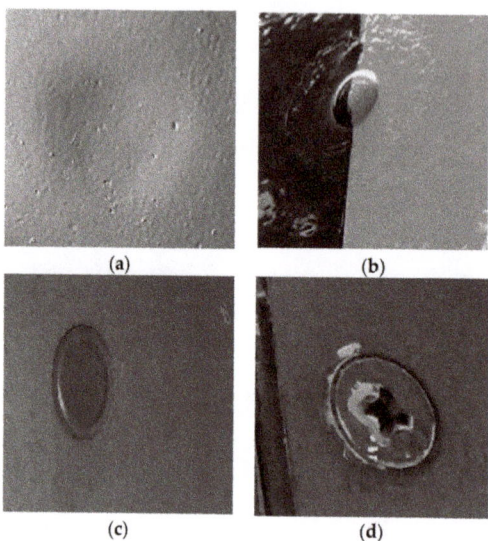

Figure 2. Sample images from the two datasets created for training the classifiers. (**a**) An image of a dent, (**b**) a lighting strike fastener repair; (**c,d**) are images with objects that are not defects.

Following the above procedure, two datasets were created, one containing images from each category of the defects described above and one contains images with and without defects. The second dataset in the no-defect category has images of screws, gaps, small plates etc., objects that the classifier will need to distinguish from the defects. Figure 2 shows images from the two datasets with and without defects.

The defect/no defect dataset, which we will refer as binary for simplicity, contains 1059 images, 576 of defects and 483 of non-defects. The other dataset, referred as the defect dataset, contains 576 images of the 7 types of defects. Both datasets were relatively small but gathering images was very challenging under the current circumstances (COVID-19 restrictions, flights reductions etc.). To try to overcome this, we carried out a custom split of the images between training, and validation, with 88% for training, 9% for validation and the rest for testing for both datasets. This was carried out to give the opportunity to the classifiers to learn as much as possible from the dataset. For the binary dataset, the splitting can be observed in Table 1 and for the defect dataset in Table 2.

Table 1. Dataset split for training, validating and testing the defect/non defect classifier.

Dataset Split	Non-Defect	Defect
Training	426	576
Validation	46	63
Testing	11	22

Table 2. Dataset split for training, validating and testing the defect classifier.

Dataset Categories	Training	Validation	Testing
Missing or Damaged Exterior Paint and Primer	77	8	3
Dents	151	25	6
Reinforcing Patch Repairs	109	10	4
Nicks, Scratches and Gouges	57	6	3
Blend/Rework Repairs	82	10	3
Lighting Strike Damage	4	1	1
Lighting Strike Fastener Repairs	11	3	2

3. Defect Classification Algorithms

As previously mentioned, one of the challenges of the classification problems in applications in aerospace is the small amount of data available. In this paper, we tried to address this by proposing a two-step classification approach with a combination of different classifiers. In the first step, a classifier decides if the image contains a defect and if this is true in the second step, the defect is classified by a different classifier. The classifiers are a combination of pretrained CNNs on ImageNet [29], which we retrained with the use of transfer learning [30]. In the first step, a DenseNet201 model is used and in the second, an ensemble of different CNNs as can be observed in Figure 3.

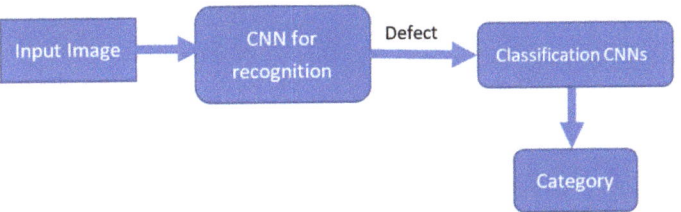

Figure 3. Block diagram of the two-step process for defect recognition and classification.

Transfer learning refers to a technique of retraining a CNN that has already been trained in very large dataset, such as Imagenet [29]. Even though the dataset that the CNN is been initially trained in is irrelevant to the problem research, ref. [30] has shown that the benefits of this technique are substantial. There are mainly two approaches on how to implement transfer learning; in the first, only the convolutional layers of the trained network are used as feature extractors [31] and then the features are fed to a different classifier, such as support vector machines [31]. In the second approach, which is used in this paper, the head of the neural network (fully connected layers) is replaced. The output of the new connected layers will match the number of the categories of our classifier. The new neural network is initially trained by keeping all the weights of the convolution layers frozen/non trainable. Then, to fine tune the model, a number of the layers are unfrozen and the training of the network is updated. The basic rule for unfreezing layers is, the less the data, the less layers to unfreeze. In addition, because the initial/bottom layers of a CNN extract more abstract features that can be used in any type of image, we unfreeze (for training) the layers closer to the top of the network. Another point that needs attention during both training rounds is not to update the weights of the batch normalization layers. These layers contain two non-trainable weights, tracking the mean and variance of the inputs that usually get updated during training. So, if we unfreeze these layers during fine tuning, the updates applied will destroy what the model has learned.

The models were implemented using TensorFlow [32], as this is a well-established deep learning library, widely used for both commercial applications and research. Because TensorFlow contains around forty pre-trained networks, we needed to identify those that fit better on our datasets. To achieve this, we trained each network for five epochs with the convolutional layers frozen. To continue with fine tuning, we chose the best four pretrained networks for each classifier. For the binary classifier, the models that performed better were Mobilenet, DenseNet201, ResNet15V2 and InceptionResNetV2. For the defect classifier, the four models with the best results were EfficientNetB1, EfficientNetB5, EfficientNetB4 and DenseNet169.

To improve the performance of the chosen models, we fine-tuned them for another ten epochs. For fine-tuning, we unfroze the last 10% of the layers of each model and reduced the learning rate by a factor of ten compared to the initial one. In addition, techniques of reduce learning and early stopping were used. Both techniques are included in TensorFlow libraries. In the reduce learning technique, the learning rate of the optimizer is reduced if the validation loss has not improved for a certain number of epochs. Similar in the early stopping as the name suggests, training stops if our metric (in this case, validation loss) has not improved for a certain number of epochs and the graph with the best weights is saved. Both techniques were used to prevent overfitting.

In addition to the CNN, a random forest was trained. The initial idea was to use in the first step both the CNN and the random forest but the overall benefit of this was low. For training, the random forest we have extracted the features of Hu moments, color histogram and Haralick. The overall accuracy of the random forest classifier was 76%.

4. Results

As discussed in the previous chapter, the initial training of five epochs has been carried out for each of the pretrained models of TensorFlow. The results of the four best networks for the defect recognition can be observed in Table 3.

Table 3. Performance of the 4 best out of 40 pretrained networks for the binary classifier after 5 epochs.

Model	Validation Accuracy	Testing Accuracy
Mobilenet	0.80	0.63
DenseNet201	0.84	0.81
ResNet152V2	0.74	0.88
InceptionResNetV2	0.79	0.85

The results of the best four networks for the defect classification can be observed in Table 4.

Table 4. Performance of the 4 best out of 40 pretrained networks for the defect classifier after 5 epochs.

Model	Validation Accuracy	Testing Accuracy
EfficientNetB1	0.60	0.68
EfficientNetB5	0.63	0.68
EfficientNetB4	0.71	0.63
DenseNet169	0.70	0.60

As expected, due to the small number of images and due to the lack of fine tuning especially for the defect classifier, the accuracy in both the validation and testing images was relatively low. At this stage, no further analysis was carried out or any extra metrics, such as confusion matrices or classification reports, as the purpose was to identify the best CNNs for each of the datasets.

After this initial training, each of the networks were further trained, as discussed for another ten epochs. The results of the training can be observed in Table 5.

Table 5. Performance of the 4 best pretrained networks for binary classifier after fine tuning for a total of 15 epochs.

Model	Validation Loss	Validation Accuracy	Testing Accuracy
MobileNet	0.39	0.79	0.63
DenseNet201	0.46	0.84	0.82
InceptionResNetV2	0.43	0.77	0.69
ResNet152V2	0.61	0.78	0.66

The same procedure was followed for the other set of classifiers for the defect classification. The results can be observed in Table 6.

Table 6. Performance of the 4 best pretrained networks for defect classifier after fine tuning for a total of 15 epochs.

Model	Validation Loss	Validation Accuracy	Testing Accuracy
EfficientNetB1	0.76	0.66	0.72
EfficientNetB5	0.52	0.85	0.82
EfficientNetB4	0.54	0.79	0.72
DenseNet169	0.82	0.71	0.82

To understand better the behavior of the CNNs while trained, the validation loss was taken into account. This metric, together with the validation accuracy, can illustrate when the CNN will start overfitting. Usually, when the validation loss does not improve, but the validation accuracy does, overfitting occurs. This is also the main reason why we used the techniques of reduce learning and early stopping.

To decide which of the above eight CNNs to use in the proposed system, further metrics were produced. For each of the models, a classification report and a confusion matrix was produced to measure the performance in the test data. A classification report measures the values of precision, recall and F1-score [33]. Precision quantifies the number of correct positive predictions. It is defined as the ratio of true positives divided by the sum of true positives and false positives [33]. It shows how precise/accurate the model is. It is very useful if the false positive cost is high, which in our case was not. If one misclassifies a non-defect, it will produce an extra load of work for the inspector but it is not critical. Recall is the ratio of correctly predicted positive predictions against all the predictions in the actual class [33]. It is the ratio of true positives divided by the sum of true positives and false negatives. In simple terms, recall shows how many of the predictions in the class are actual positives. It is the metric we can use if there is a high cost of false negatives; in our case, if we misclassify a defect as non-defect. The F1 score is calculated as the multiplication of precision and recall, divided by the sum of precision and recall and then multiplied by 2 [33]. The F1 score can be interpreted as the harmonic mean of both precision and recall. The F1 score can also be interpreted as the average of precision and recall. It is a very valuable metric, especially when both errors caused by false positives and false negatives are undesirable.

Taking into consideration all the above, we created a classification report with the above metrics for each of the models.

In Table 7, the combined classification reports can be observed for all four models for defect recognition and in Table 8, the combined confusion matrices.

From the above tables, we can observe that DenseNet201 performs very well with high precision. The results from the confusion matrix show that the model has predicted correct eighteen out of the twenty-two images containing a defect and nine out of eleven images for the no defect category.

Comparing InceptionResNetV2 and DenseNet201, we can observe that the first has a better precision than DenseNet201 for the defect category by its recall value being much lower. This is also reflected in the confusion matrix, where InceptionResNetV2 has more false negatives. In addition, the F1 score for DenseNet201 is higher in both categories. Because misclassifying a defect is critical in our application, we can state that DenseNet201 performs better.

From the above results, in can be observed that DenseNet201 has the best overall accuracy with 81.82%, the best precision and recall values for the defect class. In addition, it has the least false negatives and the best F1 score for both classes. Another test we performed was to combine the classifiers in an ensemble to investigate whether any improvements in the metrics were possible. The ensemble of classifiers did not give better results, compared to DenseNet201.

Table 7. Combined classification reports for defect recognition classifiers.

	MobileNet			
	Precision	Recall	F1 Score	Sum of Images
Defect	0.83	0.68	0.75	22
No Defect	0.53	0.72	0.61	11
Accuracy				69.70%
	ResNet15V2			
	Precision	Recall	F1 Score	Sum of Images
Defect	0.88	0.68	0.76	22
No Defect	0.56	0.81	0.66	11
Accuracy				72.73%
	InceptionResNetV2			
	Precision	Recall	F1 Score	Sum of Images
Defect	0.93	0.68	0.78	22
No Defect	0.58	0.90	0.71	11
Accuracy				75.76%
	DenseNet201			
	Precision	Recall	F1 Score	Sum of Images
Defect	0.9	0.82	0.85	22
No Defect	0.69	0.82	0.75	11
Accuracy				81.82%

Table 8. Combined confusion matrices for defect recognition classifiers.

	MobileNet	
Actual	Predicted Class	Predicted Class
	Defect	No Defect
Defect	15	7
No Defect	3	8
	ResNet15V2	
Actual	Predicted Class	Predicted Class
	Defect	No Defect
Defect	15	7
No Defect	2	9
	InceptionResNetV2	
Actual	Predicted Class	Predicted Class
	Defect	No Defect
Defect	15	7
No Defect	1	10
Actual	Predicted Class	Predicted Class
	Defect	No Defect
Defect	18	4
No Defect	2	9

The same procedure was followed for the defect classification models and the results of the metrics and confusion matrices can be observed in Tables 9–16.

Table 9. Classification report of Dense169 for defect recognition.

	Dense169			
	Precision	Recall	F1 Score	Sum of Images
Missing or Damaged Exterior Paint and Primer	0.22	0.66	0.33	3
Dents	0.67	0.33	0.44	6
Reinforcing Patch Repairs	1	0.5	0.66	4
Nicks, Scratches and Gouges	1	0.33	0.5	3
Blend/Rework Repairs	0.5	0.66	0.57	3
Lighting Strike Damage	1	1	1	1
Lighting Strike Fast Repairs	1	1	1	2
Accuracy				54.55%

Table 10. Confusion natrix for Dense 169.

Actual	Predicted Class						
	Missing/Damaged Exterior Paint and Primer	Dents	Reinforcing Patch Repairs	Nicks, Scratches and Gouges	Blend/Rework Repairs	Lighting Strike	Lighting Strike Fast Repairs
Missing/Damaged Paint and Primer	2	1	0	0	0	0	0
Dents	3	2	0	0	1	0	0
Reinforcing Patch Repairs	2	0	2	0	0	0	0
Nicks, Scratches and Gouges	1	0	0	1	1	0	0
Blend/Rework Repairs	1	0	0	0	2	0	0
Lighting Strike	0	0	0	0	0	1	0
Lighting Strike Fast Repairs	0	0	0	0	0	0	2

Table 11. Classification report of EfficientNetB1 for defect classification.

	EfficientNetB1			
	Precision	Recall	F1 Score	Sum of Images
Missing or Damaged Exterior Paint and Primer	0.6	1	0.75	3
Dents	1	1	1	6
Reinforcing Patch Repairs	0.5	0.5	0.5	4
Nicks, Scratches and Gouges	0	0	0	3
Blend/Rework Repairs	0.66	0.66	0.66	3
Lighting Strike Damage	1	1	1	1
Lighting Strike Fast Repairs	1	1	1	2
Accuracy				72.73%

Table 12. Confusion matrix of EfficientNetB1.

Actual	Predicted Class						
	Missing/Damaged Exterior Paint and Primer	Dents	Reinforcing Patch Repairs	Nicks, Scratches and Gouges	Blend/Rework Repairs	Lighting Strike	Lighting Strike Fast Repairs
Missing or Damaged Exterior Paint and Primer	3	0	0	0	0	0	0
Dents	0	6	0	0	0	0	0
Reinforcing Patch Repairs	1	0	2	1	0	0	0
Nicks, Scratches and Gouges	1	0	1	0	1	0	0
Blend/Rework Repairs	0	0	1	0	2	0	0
Lighting Strike Damage	0	0	0	0	0	1	0
Lighting Strike Fast Repairs	0	0	0	0	0	0	2

Table 13. Classification report of EfficientNetB4 for defect classification.

	EfficientNetB4			
	Precision	Recall	F1 Score	Sum of Images
Missing or Damaged Exterior Paint and Primer	0.5	1	0.66	3
Dents	0.83	0.83	0.83	6
Reinforcing Patch Repairs	0.5	0.5	0.5	4
Nicks, Scratches and Gouges	1	0.33	0.5	3
Blend/Rework Repairs	0	0	0	3
Lighting Strike Damage	1	1	1	1
Lighting Strike Fast Repairs	1	1	1	2
Accuracy				63.64%

Table 14. Confusion matrix of EfficientNetB4.

Actual	Predicted Class						
	Missing/Damaged Exterior Paint and Primer	Dents	Reinforcing Patch Repairs	Nicks, Scratches and Gouges	Blend/Rework Repairs	Lighting Strike	Lighting Strike Fast Repairs
Missing or Damaged Exterior Paint and Primer	3	0	0	0	0	0	0
Dents	1	5	0	0	0	0	0
Reinforcing Patch Repairs	0	1	2	0	1	0	0
Nicks, Scratches and Gouges	1	0	0	1	1	0	0
Blend/Rework Repairs	1	0	2	0	0	0	0
Lighting Strike Damage	0	0	0	0	0	1	0
Lighting Strike Fast Repairs	0	0	0	0	0	0	2

Table 15. Classification report of EfficientNetB5 for defect classification.

	EfficientNetB5			
	Precision	Recall	F1 Score	Sum of Images
Missing or Damaged Exterior Paint and Primer	1	1	1	3
Dents	1	0.83	0.90	6
Reinforcing Patch Repairs	0.16	0.25	0.2	4
Nicks, Scratches and Gouges	1	0.66	0.8	3
Blend/Rework Repairs	0	0	0	3
Lighting Strike Damage	1	1	1	1
Lighting Strike Fast Repairs	1	1	1	2
Accuracy				63.64%

Table 16. Confusion matrix of EfficientNetB5.

Actual	Predicted Class						
	Missing/Damaged Exterior Paint and Primer	Dents	Reinforcing Patch Repairs	Nicks, Scratches and Gouges	Blend/Rework Repairs	Lighting Strike	Lighting Strike Fast Repairs
Missing or Damaged Exterior Paint and Primer	3	0	0	0	0	0	0
Dents	0	5	1	0	0	0	0
Reinforcing Patch Repairs	0	0	1	0	3	0	0
Nicks, Scratches and Gouges	0	0	1	2	0	0	0
Blend/Rework Repairs	0	0	3	0	0	0	0
Lighting Strike Damage	0	0	0	0	0	1	0
Lighting Strike Fast Repairs	0	0	0	0	0	0	2

From the above matrices, the performance of the models for the defect classification is relatively low. However, this is due to the number of images in the dataset and because the dataset was unbalanced. To improve performance and ensure the predictions are more consistent, we used the ensemble model. We combined all four models to create a new model in which the input image is fed into all four models. The predictions of each of the models are passed to a layer that is added at the end of the model. This final layer averages the predictions of the four models and returns array with the new values. This technique, especially in our case where the performance of the models is similar, provides a more consistent outcome for all the different classes. The results for the ensemble model can be observed in Tables 17 and 18.

Table 17. Classification report of the ensemble model for defect classification.

	Ensemble			
	Precision	Recall	F1 Score	Sum of Images
Missing or Damaged Exterior Paint and Primer	0.6	1	0.75	3
Dents	1	0.83	0.90	6
Reinforcing Patch Repairs	0.5	0.5	0.5	4
Nicks, Scratches and Gouges	0.5	0.33	0.4	3
Blend/Rework Repairs	0.33	0.33	0.33	3
Lighting Strike Damage	1	1	1	1
Lighting Strike Fast Repairs	1	1	1	2
Accuracy				**68.18%**

Table 18. Confusion matrix of the Ensemble.

Actual	Predicted Class						
	Missing/Damaged Exterior Paint and Primer	Dents	Reinforcing Patch Repairs	Nicks, Scratches and Gouges	Blend/Rework Repairs	Lighting Strike	Lighting Strike Fast Repairs
Missing or Damaged Exterior Paint and Primer	3	0	0	0	0	0	0
Dents	1	5	0	0	0	0	0
Reinforcing Patch Repairs	0	0	2	1	1	0	0
Nicks, Scratches and Gouges	1	0	0	1	1	0	0
Blend/Rework Repairs	0	0	2	0	1	0	0
Lighting Strike Damage	0	0	0	0	0	1	0
Lighting Strike Fast Repairs	0	0	0	0	0	0	2

For the ensemble model, although in some categories it may have worse performance than others, its overall performance is better. It has positive predictions for all the categories in comparison with other models and its overall accuracy is above the average value of the models.

Finally, we tested the whole pipeline of our algorithm. We first fed the test images to the defect recognition model and then, if the image had a defect, we passed it to the defect classifier. As a defect recognition model, we have chosen the DenseNet201 and for the defect classification, the ensemble model. As we have used the same test dataset, the results of the defect recognition model are the same as Tables 7 and 8 and for the ensemble, similar to the Tables 17 and 18. However, by filtering through the first step, the images that we achieved 100% accuracy for were the categories of the missing or damaged exterior paint and primer and dents.

Although the results are promising, the overall accuracy of the defect classifier is low. As previously mentioned, this is mainly due to the small number of images and because the dataset is very unbalanced. Taking into consideration the accuracy for the defect recognition classifier together with the number of images, we believe that by having around five hundred images for each defect category, we will be able to improve significantly not only the performance of the defect classifier but also of the overall process.

5. Conclusions

In this paper, we have presented the development of a two-step process for defect recognition and classification of aircraft structures. A dataset was created from real aircraft defects taken in TUI's maintenance hangar. On the one hand, the lack of defects on aircrafts made the creation of the dataset very challenging and on the other, the recognition of defects is crucial for the safety of the passengers and crew. To overcome this, we proposed a two-step process method. Firstly, we recognized the defect and then we classified it. This method has the advantage of using two different classifiers, one for defect recognition and one for defect classification. By splitting the process of defect recognition and classification in two, we improved the accuracy. This is because first, we can train the defect recognition model with more data, thus making it more accurate. In addition, in this first step, we perform with higher accuracy the most significant part of finding the defect. Secondly, we use a dedicated classifier for defect classification. This gives the opportunity to the second classifier to learn more effectively the differences between the different types of defects, as it does not have to learn any of the non-defect images.

The results of the first step had an accuracy **81.82%**, which is quite high considering the small training dataset. In the second step, for the defects of missing or damaged exterior paint and primer and dents, we achieved 100% accuracy.

Although the results are promising, future work will be carried out in increasing the defect dataset, especially in adding more images in the very small categories to improve the unbalanced dataset. In addition, the process will be combined with a UAV inspection for real time recognition and classification

Author Contributions: Conceptualization, N.P.A. and A.T.; Methodology, All Authors.; Software, N.P.A.; Validation, M.D.; Resources, All Authors; Writing—Original Draft reparation, N.P.A.; Writing—Review and Editing, N.P.A., A.T. and M.D.; Visualization, N.P.A.; Supervision, N.P.A. and A.T. All authors have read and agreed to the published version of the manuscript.

Funding: This research was supported and funded by the British Engineering and Physics Sciences Research Council (EPSRC IAA project).

Institutional Review Board Statement: Not applicable.

Informed Consent Statement: Not applicable.

Conflicts of Interest: The authors declare no conflict of interest.

References

1. The EU in the World, Eurostat 2020 Edition. 2020. Available online: https://ec.europa.eu/eurostat/web/products-statistical-books/-/ks-ex-20-001 (accessed on 11 April 2022).
2. The EU in the World, Eurostat 2021 Edition. 2021. Available online: https://ec.europa.eu/eurostat/web/products-statistical-books/-/ks-ei-21-001 (accessed on 11 April 2022).
3. Sprong, J.; Jiang, X.; Polinder, H. Deployment of Prognostics to Optimize Aircraft Maintenance—A Literature Review: A Literature Review. In Proceedings of the Annual Conference of the PHM Society, Scottsdale, AZ, USA, 21–26 September 2019. [CrossRef]
4. Gunatilake, P.; Siegel, M.; Jordan, A.G.; Podnar, G.W. Image enhancement and understanding for remote visual inspection of aircraft surface. In *Nondestructive Evaluation of Aging Aircraft, Airports, and Aerospace Hardware*; SPIE: Scottsdale, AZ, USA, 1996; Volume 2945. [CrossRef]
5. Jovančević, I.; Larnier, S.; Orteu, J.-J.; Sentenac, T. Automated exterior inspection of an aircraft with a pan-tilt-zoom camera mounted on a mobile robot. *J. Electron. Imaging* **2015**, *24*, 61110. [CrossRef]
6. Gunatilake, P.; Siegel, M.; Jordan, A.G.; Podnar, G.W. Image understanding algorithms for remote visual inspection of aircraft surfaces. In *Machine Vision Applications in Industrial Inspection V*; SPIE: San Jose, CA, USA, 1997; Volume 3029, pp. 2–14. [CrossRef]
7. Chu, B.; Jung, K.; Han, C.-S.; Hong, D. A survey of climbing robots: Locomotion and adhesion. *Int. J. Precis. Eng. Manuf.* **2010**, *11*, 633–647. [CrossRef]
8. Zhiwei, X.; Muhua, C.; Qingji, G. The structure and defects recognition algorithm of an aircraft surface defects inspection robot. In *2009 International Conference on Information and Automation, Zhuhai, China, 22–25 June 2009*; IEEE: Zhuhai, China; Macau, China, 2009; pp. 740–745. [CrossRef]
9. Lal Tummala, R.; Mukherjee, R.; Xi, N.; Aslam, D.; Dulimarta, H.; Xiao, J.; Minor, M.; Dang, G. Climbing the walls [robots]. *IEEE Robot. Autom. Mag.* **2002**, *9*, 10–19. [CrossRef]
10. A. F. R. Laboratory. Robotic arm tool poised to save costly inspection time. *Int. J. Precis. Eng. Manuf.* 2018. Available online: https://www.afspc.af.mil/News/Article-Display/Article/1088209/robotic-arm-tool-poised-to-save-costly-inspection-time/ (accessed on 11 April 2022).
11. Nansai, S.; Mohan, R.E. A survey of wall climbing robots: Recent advances and challenges. *Robotics* **2016**, *5*, 14. [CrossRef]
12. Addabbo, P.; Angrisano, A.; Bernardi, M.L.; Gagliarde, G.; Mennella, A.; Nisi, M.; Ullo, S. A UAV infrared measurement approach for defect detection in photovoltaic plants. In Proceedings of the 2017 IEEE International Workshop on Metrology for AeroSpace (MetroAeroSpace), Padua, Italy, 21–23 June 2017; pp. 345–350.

13. Malandrakis, K.; Dixon, R.; Savvaris, A.; Tsourdos, A. Design and Development of a Novel Spherical UAV. *IFAC-PapersOnLine* **2016**, *49*, 320–325. [CrossRef]
14. Morgenthal, G.; Hallermann, N. Quality Assessment of Unmanned Aerial Vehicle (UAV) Based Visual Inspection of Structures. *Adv. Struct. Eng.* **2014**, *17*, 289–302. [CrossRef]
15. Malandrakis, K.; Savvaris, A.; Domingo, J.A.G.; Avdelidis, N.; Tsilivis, P.; Plumacker, F.; Zanotti Fragonara, L.; Tsourdos, A. Inspection of Aircraft Wing Panels Using Unmanned Aerial Vehicles. In Proceedings of the 2018 5th IEEE International Workshop on Metrology for AeroSpace (MetroAeroSpace), Rome, Italy, 20–22 June 2018; pp. 56–61.
16. Mumtaz, R.; Mumtaz, M.; Mansoor, A.B.; Masood, H. Computer aided visual inspection of aircraft surfaces. *Int. J. Image Processing* **2012**, *6*, 38–53.
17. Ortiz, A.; Bonnin-Pascual, F.; Garcia-Fidalgo, E.; Company-Corcoles, J. Vision-based corrosion detection assisted by a micro-aerial vehicle in a vessel inspection application. *Sensors* **2016**, *16*, 2118. [CrossRef] [PubMed]
18. Tzitzilonis, V.; Malandrakis, K.; Zanotti Fragonara, L.; Gonzalez Domingo, J.A.; Avdelidis, N.P.; Tsourdos, A.; Forster, K. Inspection of Aircraft Wing Panels Using Unmanned Aerial Vehicles. *Sensors* **2019**, *19*, 1824. [CrossRef] [PubMed]
19. Li, Y.; Huang, H.; Xie, Q.; Yao, L.; Chen, Q. Research on a surface defect detection algorithm based on MobileNet-SSD. *Appl. Sci.* **2018**, *8*, 1678. [CrossRef]
20. Malekzadeh, T.; Abdollahzadeh, M.; Nejati, H.; Cheung, N.-M. Aircraft fuselage defect detection using deep neural networks. *arXiv* **2017**, arXiv:1712.09213.
21. Tao, X.; Zhang, D.; Ma, W.; Liu, X.; Xu, D. Automatic metallic surface defect detection and recognition with convolutional neural networks. *Appl. Sci.* **2018**, *8*, 1575. [CrossRef]
22. Cha, Y.-J.; Choi, W.; Suh, G.; Mahmoudkhani, S.; Büyüköztürk, O. Autonomous structural visual inspection using region-based deep learning for detecting multiple damage types. *Comput. Aided Civ. Infrastruct. Eng.* **2018**, *33*, 731–747. [CrossRef]
23. Kim, B.; Cho, S. Automated Vision-Based Detection of Cracks on Concrete Surfaces Using a Deep Learning Technique. *Sensors* **2018**, *18*, 3452. [CrossRef] [PubMed]
24. Zhang, R.; Wang, Z.; Zhang, Y. Astronaut visual tracking of flying assistant robot in space station based on deep learning and probabilistic model. *Int. J. Aerosp. Eng.* **2018**, *2018*, 6357185. [CrossRef]
25. Cha, Y.-J.; Choi, W. Vision-based concrete crack detection using a convolutional neural network. In *Dynamics of Civil Structures*; Conference Proceedings of the Society for Experimental Mechanics Series; Caicedo, J., Pakzad, S., Eds.; Springer International Publishing: Cham, Switzerland, 2017; Volume 2, pp. 71–73. [CrossRef]
26. Kang, D.; Cha, Y.-J. Damage detection with an autonomous UAV using deep learning. In *Sensors and Smart Structures Technologies for Civil, Mechanical, and Aerospace Systems 2018*; SPIE: Denver, CO, USA, 2018; Volume 10598. [CrossRef]
27. Kang, D.; Cha, Y.-J. Autonomous UAVs for structural health monitoring using deep learning and an ultrasonic beacon system with geo-tagging. *Comput. Aided Civ. Infrastruct. Eng.* **2018**, *33*, 885–902. [CrossRef]
28. TUI©. Available online: https://www.tuigroup.com/en-en (accessed on 11 April 2022).
29. Deng, J.; Dong, W.; Socher, R.; Li, L.-J.; Li, K.; Fei-Fei, L. ImageNet: A large-scale hierarchical image database. In Proceedings of the 2009 IEEE Conference on Computer Vision and Pattern Recognition, Miami, FL, USA, 20–25 June 2009; IEEE: Piscataway, NJ, USA, 2009; pp. 248–255.
30. Tan, C.; Sun, F.; Kong, T.; Zhang, W.; Yang, C.; Liu, C. A Survey on Deep Transfer Learning. In *The 27th International Conference on Artificial Neural Networks (ICANN 2018), Rhodes, Greece, 4–7 October 2018*; Springer: Cham, Switzerland, 2018. [CrossRef]
31. Keiller, N.; Otavio, P.; Jefersson, S. Towards better exploiting convolutional neural networks for remote sensing scene classification. *Pattern Recognit.* **2017**, *61*, 539–556. [CrossRef]
32. Abadi, M. TensorFlow: Large-Scale Machine Learning on Heterogeneous Systems. 2015. Available online: https://www.tensorflow.org (accessed on 14 April 2022).
33. Pedregosa, F.; Varoquaux, G.; Gramfort, A.; Michel, V.; Thirion, B.; Grisel, O.; Blondel, M.; Prettenhofer, P.; Weiss, R.; Dubourg, V.; et al. Scikit-learn: Machine Learning in Python. *J. Mach. Learn. Res.* **2011**, *12*, 2825–2830.

MDPI
St. Alban-Anlage 66
4052 Basel
Switzerland
Tel. +41 61 683 77 34
Fax +41 61 302 89 18
www.mdpi.com

Sensors Editorial Office
E-mail: sensors@mdpi.com
www.mdpi.com/journal/sensors

www.ingramcontent.com/pod-product-compliance
Lightning Source LLC
LaVergne TN
LVHW070726100526
838202LV00013B/1179